Where Water Flows

Where Water Flows

The Rivers of Arizona

by Lawrence Clark Powell

with photographs by Michael Collier and a Foreword by Bruce Babbitt

NORTHLAND PRESS / FLAGSTAFF, ARIZONA / 1980

FRONT COVER: *The gorge of the Little Colorado River*

FRONTIS: *A tributary creek on the San Pedro*

To John Paul Schaefer
administrator   artist   friend

A long time ago — no one knows just when — people paused beside
the running water of the desert river. Their pause became
a short stay. A short stay became a home. A home became many homes
and these became a village. Water courses were their highways.
Permanent water is the magnet that draws life. Animals
to hunt and plants to gather were plentiful. Farming and crops of
corn, teparies, squashes and other native foods were a
good possibility. And thus it was that life evolved along the river.

Bernard L. Fontana, writing
of the Río de la Santa Cruz at Bac

Contents

Foreword

AT FIRST GLANCE, *Where Water Flows: The Rivers of Arizona* is about a subject as evanescent as the Seven Golden Cities of Cíbola. Arizona has lots of rivers — mostly dry. Edwin Corle once described a thirsty pioneer who after enduring the dry rivers and mirages of southern Arizona finally reached the banks of the Colorado and remarked in genuine surprise, "Why, it's wet."

Larry Powell's thesis is that the wet rivers take on great significance precisely because they are so few. In the Sonoran desert the wet rivers are lifelines that delimit where people settle, how they live, what they grow, and ultimately, the shape of their culture. It was true for the Hohokam a thousand years ago, and it is true today. We live in an oasis culture; if and when the pumps go dry we will all return to the desert's "basic laws of abstinence and endurance."

Powell's narrative takes the rivers and their cultures in the chronology of their discovery and settlement by Europeans. He enters the San Pedro Valley in 1540 with the soldiers of the Spanish empire, winds down the Santa Cruz through the heroic age of Spanish missionaries, and finally moves up the Mogollon Rim to the more recent Anglo-American settlements in northern Arizona. Along the way we are treated to a rich brew of the history, geography, literature, and lore of three cultures: Indian, Spanish, and Anglo.

Ultimately, however, *Where Water Flows: The Rivers of Arizona* is much more than another catalog of factual curiosities and eccentric personalities. It is a small and rich book of prophecy, an intensely personal vision of what Arizona ought to be. Powell's vision of Arizona rings truest when he pauses to visit with friends such as Eulalia Bourne at her Copper Creek homestead, and George and Holly Pilling at the Nature Conservancy's Canelo Hills Cienega ("I found people living on a harder, richer, truer level than most"). The author likes bookish cowpunchers, artisans and artists living on the land, footloose geologists, women pioneers — most anyone who has a sense of the land and its past.

He has no such affection for Phoenix and Tucson, the two great cities that have dried up and killed the Río Santa Cruz and the Río Salado. In fact, Powell is something of a latter-day Jeremiah, prophesying doom and disaster for those who live out of harmony with the desert, squandering and misusing our limited water resources.

This book could hardly appear at a more relevant time. In February the jet streams over the central Pacific plunged south, pushing one rainstorm after another across Arizona. The rushing waters overflowed dams, ripped out bridges like tinker toys, ate away farms, and flooded houses throughout the state.

The flooding Salt River sliced Phoenix in two. Angry drivers stacked up on Central Avenue in lines four to five miles long, waiting as many hours to cross. The Red Cross rushed in to provide sandwiches. Enterprising teenagers offered their services at one dollar an hour to tend cars while weary drivers headed for the corner hamburger stand. Bumper stickers and tee shirts proclaiming "I crossed the Salt" blossomed.

A few weeks later the sun was shining again, and it was business as usual. Arizona's political leaders trekked back to Washington to make our annual plea for the Central Arizona Project, an intricate and costly aqueduct system to bring up more water from the Colorado River to quench the thirst of Phoenix and Tucson. Incredulous Congressmen, aware of our urgent requests for more flood control, wondered aloud whether we weren't seeking to carry coals to Newcastle. Such are the vagaries of the desert: feast and famine, flood and drouth.

Larry Powell and I are desert neighbors. We breathe the same air, travel the same landscape, and read and reflect on many of the same books. But our work often goes in different directions. I spend my days planning for more growth, trying (without really knowing whether it is even possible) to accommodate the inevitable flow of more people without destroying the values that brought us here in the first place. In a few weeks I will call the Legislature into Special Session to enact a groundwater conservation code, a difficult and controversial task that has been put off for nearly fifty years.

On the other hand, Larry counsels that we should spend more time examining the premise that growth is always desirable and that we ought to read the lessons of the past more reflectively and be cautious about manipulating nature with too many dams and water projects. I don't know which view will ultimately prevail, but this book ought to be read by anyone who cares about finding the answer.

BRUCE BABBITT

Where Water Flows

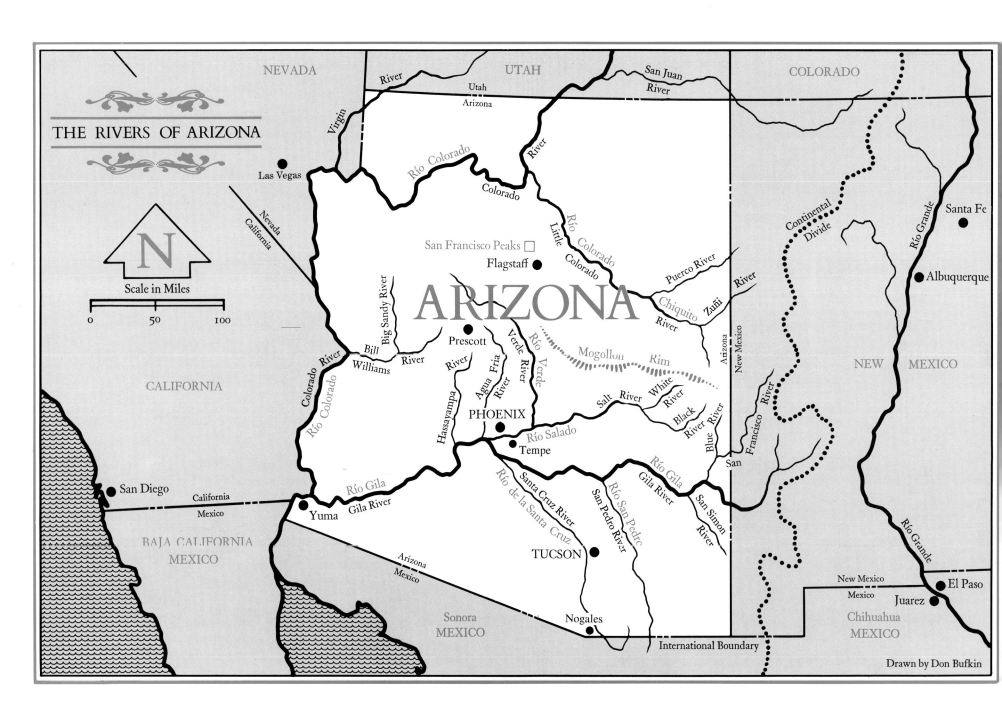

THE RIVERS OF ARIZONA

N

Scale in Miles

0   50   100

NEVADA

UTAH

Utah
Arizona

San Juan
River

COLORADO

Río Colorado

River

Virgin

River

Las Vegas

Nevada
California

Colorado

Little

Río Colorado

Continental
Divide

Santa Fe

San Francisco Peaks

Flagstaff

Colorado

Puerco River

Zuñi

River

Albuquerque

ARIZONA

Chiquito

River

CALIFORNIA

Colorado River

Río Colorado

Big Sandy River

Prescott

Bill
Williams

River

River

Hassayampa

Agua Fria
River

Verde River

Río Verde

Mogollon   Rim

Arizona

New Mexico

NEW   MEXICO

Río Grande

PHOENIX

Salt   River

White
River

Black

River   River

San Francisco River

Blue River

Río Salado

Tempe

Río Gila

Gila River

Río Gila

San

San Simon
River

San Diego

California

Mexico

Gila River

Yuma

Río Gila

Santa Cruz River

Río de la Santa Cruz

Río San Pedro

San Pedro River

NEW MEXICO

BAJA CALIFORNIA
MEXICO

Arizona

Mexico

TUCSON

El Paso

New Mexico

Mexico

Juarez

Sonora
MEXICO

Nogales

Chihuahua
MEXICO

Río Grande

International Boundary

Drawn by Don Bufkin

Prologue

IF THERE WERE ONE KEY TO THE SOUTHWEST'S HISTORY AND culture, it might well be the river systems of the region. Man goes where water flows, up and down the river trails and back and forth along the river arteries. River banks and valleys are ancient places of human habitation. Rivers are history's circulatory system.

A classic example of the river as destiny is the Río Grande, New Mexico's old lifeline, known to the Spaniards as *el río bravo del norte*. Because it rises in the snowy San Juan Rockies of Colorado, the Río Grande is a perennial rather than an intermittent river. For its flow, Indians settled on its shores and have never left. Along its course and up its tributaries, Spaniards colonized. Their descendants are still there in this tri-cultural state. On the banks of the Río Grande at the east-west, north-south crossroads, Albuquerque has become New Mexico's center of power, even as a perennial river-site has seen Phoenix rise to dominance in Arizona.

Can Arizona be interpreted in terms of its river system? While it is true that rivers have shaped its destiny, Arizona is geographically and climatically different from New Mexico and without a counterpart to the Río Grande. Although the Colorado is a greater river in length and volume, its effect on Arizona has been limited by its physical characteristics, as it has dug down into the northern part of the state, then widened to form the border with California. There it flows alongside the lowest, hottest, and most arid part of the state. In Arizona, the Colorado was either too deeply canyoned or too subject to flooding for early man to establish himself on its shores the way he did along the Gila.

The Gila is Arizona's other major river, snaking across the state from New Mexico to California. Its valley once held aboriginal cultures akin to those of the Río Grande, although in flow the Gila is a lesser stream. The early Gilans were good hydrologists, capable of devising canal systems to bring water to their fields and dwellings. These were situated at high points on the flood plain.

Drought, or soil exhaustion from over-irrigation, or hostile tribes ended the Gila culture. The Gilans' descendants, the Pimas, call the vanished puebloans the Hohokam, the People Who Went Away, or in a less romantic translation, Those Who Were All Used Up. Casa Grande, a ruin on the river's middle reaches left by a people known as the Saladoans, who also disappeared, is evidence of their skill as architects.

By the nineteenth century these river peoples' successors, who came from Mexico, had settled along the middle Gila as village farmers. We call them Pimas and their land, the Pimería Alta. It was a vast domain, stretching from the Río Concepción in Sonora to the Gila, and from the Río San Pedro to the Gulf of California. The first trappers, then the military, and finally the argonauts found these Pimas a friendly people. Their culture was dried up by the Anglos who moved in upstream and canalled off the river's water to their fields.

Today, Arizona's rivers are dammed to prevent flooding and to generate power, and they are also diverted for agriculture, mining, and urban needs. Their water is, as it has always been, the essential source of life in this arid land. Modern Arizonans confine their worship to churches, ignoring the gods who dwell in the springs, cienegas, and rivers. Earlier Arizonans recognized the holiness of water, and to this day the Hopis invoke rain with serpentine rituals, while the Papagos dance and sing for power over the precious element. Here is what Coronado's unknown chronicler, who came with the conquistador in 1540, wrote: "What the Indians worship most is water, to which they offer painted sticks, plumes, and powders made of yellow flowers, this usually at springs. Sometimes they offer such turquoises as they have." And Coronado himself reported to the viceroy: "The water is what these Indians worship, because they say it is what makes the corn grow and sustains their life."

Cotton, not corn, is now Arizona's thirstiest water user. If we should run out of water, could it be because the growers have not made offerings at their pumps and standpipes? The owners are often distant corporations whose bylaws make no provisions for the water gods.

For a quarter century, Arizona and the wider Southwest have been my inspiration and source as a writer. In whatever form chosen, whether history, biography, essay, or novel, my vision has been formed by the land, its configurations and climate. *Mountains determine,* I once said,

echoing Ross Calvin's *Sky Determines.* Now I say *rivers determine,* as I see water that makes rivers wearing down mountains.

Not being a professional historian, geographer, or hydrologist, even less a geomorphologist, I am a writer who has gone about his adopted land from the Yuma Crossing to the Four Corners and from the North Rim to La Noria, sensing the earth and sky and prevailing winds; and for the past ten years dwelling on the *bajada* of the Santa Catalinas, overlooking the Old Pueblo to the mountains of Mexico.

From this base I have followed the river trails, seeking their meaning in Southwest history and culture and also to our future in a land of dwindling natural resources. When the going grew rough, I traded the comfort of a Citroën for the ruggedness of a Scout, and finally I took off in a plane for a higher view of things.

This happened when fate came to my door in the person of a young photographer. "I'm Michael Collier from Flagstaff," he said, "and I want to illustrate your book." When he proved to have a flyer friend named Christopher Condit who owned a Cessna 170, we formed a team. My vision thus gained a new dimension, not the least of which came from their being graduate geologists, from Stanford and Northern Arizona. In the end it was a tradeoff of their geology and aviation for my history and literature. Our common vocabulary of expletives bridged the generation gulf.

Throughout years of commercial flights across America I had come to know the country's physical outlines. In my mind was the advice Bernard DeVoto gave to teachers of western history — to take their students aloft in order to gain a better understanding of why the westward movement went the ways it did, up and down the river valleys and through the mountain passes. This is an old truth, that man goes where water flows.

Now that I have followed most of the rivers of Arizona by land and by air, here are the fruits of these travels. They also derive from obscure memories of childhood crossings of the Colorado at Needles and Yuma.

Does this preoccupation with flowing water come from the heritage of one born on the Potomac to parents from opposite shores of the Hudson? From wherever, I say again *rivers determine*. It has taken my three score and thirteen years to arrive at this book in which the rivers of Arizona are the points of departure and return.

Underlying this gypsying was a sombre sense that Arizona is on a collision course with destiny. Those with historical knowledge of the world's deserts recognize that in the end it is fatal for man to urbanize a fragile environment. Cities with diminishing water are doomed.

Today, urban and agricultural Arizona are drawing on their water supply faster than it is being replaced. Our years as dwellers on an artificially verdant desert are numbered. Did not the poet warn us? "The wave falls and the hand falls. Thou shalt not always walk in the sun."

What is our prospect as tenants? It is this: when our lease expires, the land will go back to its original owners, the desert-dwelling Hopis, Pimas, Papagos, Mojaves, Yumas, and their kinfolk. And it will only revert to them if they are able to return to the ways of their ancestors, who lived by the desert's basic laws of abstinence and endurance.

And so in the end, this book of some history, more geography, and much travel and reflection, is a lament and a farewell. Yet it is also a thanksgiving for those fruitful years when the rains fell and the rivers ran and the desert bloomed.

Río San Pedro

BECAUSE ARIZONA LIES ON THE WESTERN SLOPE OF THE CONTINENTAL Divide, all its rivers flow toward the Pacific. Most of them run west-southwest; none to the east; three flow north-northwest. These latter are the San Pedro, the Santa Cruz, and the Colorado Chiquito, the Little Colorado.

Most of us grew up to believe that north is up and south is down, and that water always flows downhill. The latter is true except when water is siphoned, pumped, or otherwise forced to go against the law of gravity

Their flowing north does not mean that these three Arizona rivers are defying gravity. Parts of southern Arizona are higher than central Arizona, and thus the San Pedro and the Santa Cruz do indeed flow north downhill to union with the Gila. The northern slope of the White Mountains, the state's geographical backbone, descends toward the Colorado River, and so the Little Colorado, which rises on the northern shoulder of Mt. Baldy, runs downhill north-northwest to join the greater river near the head of the Grand Canyon.

The valley of the San Pedro is Arizona's oldest river trail in historic

Left: Smelter at San Manuel on the San Pedro

times. It rises in Mexico some forty miles south of the border near Cananea in Sonora. The dividing watershed is a delicate line between the Río Sonora, flowing west into the Gulf of California, and the Río San Pedro, flowing north one hundred fifty miles to the Gila at Winkelman. Folklore has it that if two Cananeos stand back to back on the courthouse steps and make water, one will replenish the San Pedro, the other the Sonora.

The first European believed to have crossed the San Pedro up near its source was that transcontinental Ulysses, Alvar Nuñez Cabeza de Vaca. Shipwrecked on the coast of Florida in 1528, this hardy Spaniard from Extremadura and a few fellow survivors took eight years to reach the West Coast on foot. His was an incredible odyssey of perils and survival. While en route, Cabeza de Vaca discovered that he had the power of healing, with the result that he gathered a following of hundreds of Indians who begged to be healed of their sicknesses. He also heard rumors of fabled golden cities. His is a wondrous story, told by him in a letter to the king upon his return to Spain. Cabeza de Vaca's narrative has been imaginatively interpreted in our time by Haniel Long, the sage of Santa Fe, in his *Interlinear to Cabeza de Vaca*.

When the viceroy heard Cabeza de Vaca's tale of rumored cities of gold, he sent Fray Marcos de Niza and party to find these new sources of wealth for the crown. The young priest had already gained a reputation as an explorer. He sent Estevaníco, a Black, on ahead to learn the mood of the Indians. Although Estevaníco succeeded in reaching Hawíkuh and Zuñi pueblos, greed and lust were his undoing. The Indians killed him. When the news was brought by the survivors to Fray Marcos, who had gotten as far as the Gila, the priest gave up and returned to Mexico. His entry into what is now Arizona is commemorated by a monument on the border at Lochiel-La Noria.

The Spaniards of that time cared more for gold than for God. In dreaming of another conquest similar to those that enriched the Spanish crown with Inca and Aztec gold, the viceroy now dispatched a company of several hundred soldiers, priests, Indians, livestock, and supplies under the command of Don Francisco Vásquez de Coronado. His order was to find and reduce the golden cities of Cíbola — this being the Mexican-Spanish name for buffalo and by extension for the grasslands where the buffalo roamed — and to bring back gold and slaves.

In the summer of 1540, Coronado and his company marched down the river trail of the San Pedro. A drive of cattle that in the beginning served for meat had either been eaten or lost. There are those who believe that Coronado introduced cattle into Arizona and that the descendants are still here. Not true. Those that the Spaniards did not eat, the Apaches probably did. Not until Eusebio Francisco Kino, the Jesuit missionary, brought them to the Southwest one hundred fifty years later did cattle and sheep come to stay, multiply, and overgraze the land.

Because the San Pedro flows in volume only after flash floods, the Coronado expedition would have used the riverbed for a roadway. The gradient is gentle. Before ruinous overgrazing by the longhorns driven from Texas in the nineteenth century to supply American troops on the Apache frontier, the river valley was lush with grama and other grasses that served to prevent erosion. Aboriginal dwellers found it a

fair place. In prehistoric times it was a domain of the mammoth, whose bones are now found in the banks of the river and its tributary creeks.

Mountain ranges wall the San Pedro from source to mouth — the Huachucas and Whetstones, Rincons and Santa Catalinas on the west; and on the east the Dragoons, Winchesters, and Galiuros, all contributing runoff to the river's flow.

At some point below Benson where Interstate 10 now crosses the river, Coronado followed Fray Marcos's route and crossed the Galiuros into Sulphur Spring Valley, then took Eagle Pass through the Pinaleños into the valley of the Gila. Bolton's fieldwork determined that Coronado crossed the river where the site of Bylas is today, then marched his men and horses onto the forested rim of the White Mountains and northeast to Zuñi, which he conquered.

Rejecting it scornfully for the mud pueblo it was, Coronado kept on until he crossed the Río Grande and reached present-day Kansas and the plains of Cíbola. As a young Nebraskan four centuries later, Willa Cather heard tales of Coronado's penetration into this heart of America. They formed in her mind as her first story of the Southwest, called "The Enchanted Bluff."

Nothing came of the Coronado "gold rush." The Southwest was dropped from the Spanish agenda for another one hundred fifty years. Yet such are the vagaries of history that the name of Coronado is the most omnipresent Spanish place name in the Southwest, from the Río Grande to the islands off San Diego. Is it because of his quixotic quest or for the music of his patronym?

Since those earliest *entradas*, the valley of the San Pedro has not served as a major river trail. The flow of peoples has been from east to west. In the beginning, few emigrants remained in Arizona. The land was too high and cold in the northern mountains, too low and hot in the southern deserts. And there were the Apaches, fanatically determined to keep settlers from their hunting lands. In vain. By 1886 the fierce tribes were penned on reservations.

The Mormons were among the first whites to settle along the San Pedro. In 1887 the greatest earthquake ever to hit southern Arizona and Sonora cracked the earth up-river from Benson, releasing a rush of artesian water. It is still flowing at the little Mormon town of St. David. Ponds and pools are shaded by towering *álamos* whose thirst is slaked by the abundant water. In the reedy backwaters, yellow-headed blackbirds break their migratory flight between Canada and Mexico.

Because it drains a smaller area than the Gila and the Colorado, and because the mountains whose waters swell the river are lower than the Rockies and those of the Gila's headwaters, the San Pedro is not subject to major floods. The worst occur when summer storms from off the West Coast or the Gulf of Mexico dump their rain on the Huachucas and the river runs high for a few days. The San Pedro is one of the few undammed rivers in Arizona. Its flow has never been copious nor constant enough to generate electrical power.

My first sight of the San Pedro came from a viewpoint near the summit of the Santa Catalinas, Tucson's *sierra madre*. The time was summer and I gazed down a mile or more on the white riverbed between dark borders of mesquite. The dominant colors were the yellow-browns of the foreground slopes and the blues of the Galiuros beyond the river. Because that was before the copper smelter came, like a dragon in Eden, to San Manuel on the river, the view was clear in all directions — even to Mt. Graham, crowning point of the Pinaleños and the second highest mountain in Arizona south of the San Francisco Peaks.

To reach the San Pedro from that viewpoint, a high-bed four-wheel drive vehicle is needed to descend a primitive roadway over fractured granite. The river is gained at Redington, which boasts only a ranch and a schoolhouse. The crossing is treacherous. Except in flood the water

is hidden beneath quicksand. There is no bridge for fifty miles between Mammoth and Benson. The river road along the eastern shore is rough and dusty, dropping in and out of countless arroyos.

Down-river from Redington, Copper and Aravaipa creeks are the major sources of water from the Galiuros. A road up Aravaipa leads to a wildlife sanctuary of the Nature Conservancy. There Edward Abbey, sardonic genius of Southwestern literature, once served as ranger.

Who could have foretold that the voice of the San Pedro would be a schoolteaching ranchwoman? Of the unlikely ways that lead to literature, few can match those taken by Eulalia "Sister" Bourne. Her first book was *Woman in Levi's*, published in 1957 by the University of Arizona Press. It has since taken its place on the short shelf of Arizona classics by women. Then came two more by Bourne, *Nine Months Is a Year* and *Ranch Schoolteacher*. Now nearing ninety, she has recently published a children's story called *Blue Colt*.

Bourne is going nearly as strong as ever, lecturing and autographing throughout Arizona and residing with a Mexican helper on her remote spread in the Galiuro roughs up Copper Creek from Mammoth. Wearing Levi's, a flashy shirt, boots, and a straw sombrero, "Sister" is an eyestopper at any gathering, whether of writers or riders. Her books have brought wide response. One elderly reader journeyed from Florida to announce that he had come to claim the woman of his dreams. It took her two weeks to get rid of him.

When she was given Pima County's most isolated accommodation school at Redington, she wearied of commuting weekends a hundred miles to her home in Tucson, and so she homesteaded in Peppersauce Canyon on the west side of the San Pedro in the lower Santa Catalinas. There she ran a few head of cattle.

Upon being assigned to a similar school in the Baboquivaris on the border near Sasabe, Bourne boarded there and made the weekend four hundred mile round trip to Peppersauce in her Model T. Later calls were to other remote schools in the Sierritas and Santa Ritas. Those were the years before unified schools were made possible by buses.

Eulalia Bourne's books are a saga of service and hardship leavened by love. Arizona's Superintendent of Public Instruction would do well to make them required reading in teacher-training curricula.

Bourne has also written a good account of the San Pedro Valley — its history, folklore, geology, archaeology, and weather. Her article appeared in *Arizona Highways,* February 1970, with photographs by Jeffrey Kurtzeman. This description of the San Pedro would be hard to better:

> This wild, dry, pitiless valley I have come to love is not beautiful. It is picturesque. Spectacular. These rough pinnacles and escarpments and metallic chimneys are at their best near sundown when the slanting light-rays brighten them with luminous glow. This is enhanced if there are cumulus clouds to splash the intervening depressions with purple or dark blue shadows. Every morning I can look across miles of westerly space and get the radiance on the blue Catalinas. When there is a red sunrise in the time of winter storms I sometimes see from the small window of my shack the blue celestial peaks capped with pink snow. In late afternoon I like to stand in the wide green fields that border the river, especially if there are white-faced cattle cropping the tender green, and see the giant cottonwoods along the field edges, the mesquite jungles across the sandy river bed, hemmed in by the pink bluffs of the alluvial mesas, and on the far horizon the tall blue and purple mountains. Our valley is not beautiful, but it compels attention with scenes of grandeur.

The valley of the San Pedro attracts strong women. There is

another as remarkable as Eulalia Bourne; books and ranching are their bond. Winifred Bundy appeared ten years ago in my class in Southwestern literature. Like Bourne, she wore Levi's, boots, and a flannel shirt. She was a wiry blonde with a strong handshake who looked like the Norwegian she is. She and her husband, Bob, an engineer at the Hughes plant in Tucson, run a working ranch down the San Pedro from the Benson crossing. Two sons and a daughter helped. When they reached college age, Win Bundy decided to further *her* education.

She commuted to campus in Tucson to take concurrent Master's degrees in library science and history. Toward the end of her second year I feared she had cracked under pressure when she told me that she wanted to be a bookseller.

"What about the ranch?" I asked, assuming that she was planning to open a shop in Tucson. "You can't leave it."

"Of course I won't," was her reply. "I'll have my shop at home."

"You're crazy," I said. "You haven't a telephone. No one could ever find your place. You'd better get a job in the Benson library. People like you are needed to bring life to libraries." Even as I spoke I realized (as she did) that the sedentary side of library work *would* have driven her crazy.

She recalled Emerson. "I'm going to build a better bookshop, specializing in Southwestern literature. People will find me."

Years passed. She was right. Her Singing Wind Bookshop (named for the ranch) is on its way to being one of the best of its kind in the West. People in the valleys of the San Pedro and the Santa Cruz, and elsewhere in Arizona, have come to depend upon Bundy for the best that is being published on the Southwest. And she now has a telephone.

The Bundys have ranch friends the length of the San Pedro from the border to the Gila, and if they weren't readers to start with, Win

Right: A tributary creek on the San Pedro

has converted them. One of the closest is Eulalia Bourne, Bundy's sister under the binding.

Every paradise has its peril. The valley of the Santa Cruz has Tucson, while the valley of the Salt has Phoenix, those urban sprawlers that began as small desert towns and now threaten to destroy their fragile environment. But the valley of the San Pedro's threat is not from a city — Benson bids fair to remain a modest river crossing. This valley's troubling presence is the fourth of Arizona's big C's: Cattle, Cotton, Climate, and Copper.

Three copper smelters are the blights of the San Pedro and beyond, one up near the source in Cananea, another in Hayden at the Gila junction, the third in mid-course at the company town of San Manuel. The smoke from their high stacks often veils the entire valley. And not only does the San Pedro suffer. When the wind is up-river from San Manuel, the smoke finds its way through Redington Pass between the Rincons and the Santa Catalinas and enters the valley of the Santa Cruz, creeping along the base of the Tanque Verdes and causing Tucsonans to blame themselves for the murk. If the wind is down-river, the smoke drifts through the pass at Oracle and seeps along the battlements of the Santa Catalinas to reach Tucson from the north.

The Mexican smelter, partly owned by U.S. interests, is an even deadlier agent of pollution. In heavy rain its waste ponds overflow into the young river. Then the San Pedro, from Cananea deep into Arizona, runs orange-colored with copper wastes. Fish and birds and other wildlife, as well as cattle, that depend upon the river for life are poisoned. This has provoked the International Boundary Commission to concern.

On one of our first flights in the Cessna, the three smelters served as landmarks. Around the stack at San Manuel, in itself beautiful as the wind creamed off the smoke, Chris circled to set up shots for Michael. "A good basketball player could dunk-bomb that stack!" Collier grinned, as below through the acrid fume the company town served as a sterile example of urban planning. "A watermelon would have more effect," Condit growled.

Consolation may be gained from a look ahead to the time when the ore deposits on the western slope above Mammoth are exhausted and the company turns the town over to ghosts. Arizona copper is no longer the biggest of the four C's. Climate is why the state is now growing so fast, attracting new manufacturing and tourism. The situation is not hopeless.

The road north over Canelo Pass descends to join O'Donnell Creek, which flows into Babocomari Creek, a tributary of the San Pedro. O'Donnell is not just another creek. Its water comes to the surface in a spring on the north shoulder of the Canelos and eventually creates a living cienega, one of the last in southern Arizona, where overgrazing and erosion have turned most of the creeks into *arroyos secos*.

In the 1960s this sixty-acre-long tract was acquired by the Nature Conservancy from the pioneer Knipe family as a wildlife sanctuary with a resident caretaker. Stephen Levine was one of the first to live there in the century-old ranch house. His book, *Planet Steward* (1974), is a sensitive journal of his custodianship.

George and Holly Pilling were the cienega's next caretakers. During their three-year tenure, we became friends. As only the house and utilities were provided, the Pillings had to furnish their own sustenance. George was a skilled custom furniture maker. His background included a Fine Arts degree from Yale, period furniture studies at Winterthur Museum, and the assistant directorship of the University of Arizona Art Museum. There he met and married Holly, a fine arts student.

In addition to their responsibilities for keeping the sanctuary's

fences mended, repelling predators — animal and human — and also serving as guides to visiting scientists and Audubon members who came to observe the many varieties of birds and mammals, the Pillings set up their individual shops in an old barn on the property, George as a furniture maker, Holly as a maker of pots and prints. Together they made Melinda and then Samuel. When cash ran low, George cut and sold oak firewood to the restaurant at the Sonoita crossroads and sometimes tended bar on Saturday night.

Not only did the Canelo sanctuary offer security to wildlife, but I also found refuge from the ardors of living in a city. There I came to share the Pillings' fare at noontime — wheat bread and butter, leek and potato soup, tuna or salmon salad with homegrown lettuces — while over tea or coffee I brought them news of town seventy miles away.

Even in summer the cienega holds water beneath the tufted surface. The creek itself never fails. In flood it is restrained by a low concrete dam. The trees are of great age and size — cottonwoods, sycamores, *nogales,* willows, and alders — with trailing vines that make deep shade. An abandoned orchard bears fruit — apple, pear, and quince. An alligator juniper is the patriarch of all.

I was sad when the Pillings moved to California. Several record cold winters found the old house too hard to heat. So they packed their things, including the oak crib made for Melinda and Samuel, and headed west in their old truck and trailer.

At the Canelo sanctuary, as at the Bundys' Singing Wind, I found people living on a harder, richer, truer level than most. As a youth I dreamed of joining a few others in a commune on the west coast of Baja California. The Depression ended that dream, and we became urban scratchers. Now it is good to see later generations following the gleam and sometimes finding it.

Left: Pollution on the San Pedro from the smelter at Cananea

11

Río de la Santa Cruz

ASK A TUCSONAN WHERE THE RIVER RISES AND HE IS APT TO answer, "*What* river?" Except in flood, the Santa Cruz is not much to see; its dry bed is littered with junk, its banks a dump-land of rubble and old tires. And yet this river and its tributary Rillito are why Tucson is an ancient site of human habitation. Along this riparian ribbon in the Sonoran desert, the Spanish missionaries of the seventeenth and eighteenth centuries first planted the cross.

It is hard for newcomers, accustomed to such permanent rivers as the Hudson or the Ohio, to perceive the importance of even the intermittent water of the Santa Cruz. They do not see that this stream's diminished flow is still Tucson's most valuable resource.

Throughout millions of years, the Santa Cruz and its drainage basin have created a subterranean reservoir from which we are taking water faster than it is being replenished. Until recently, surface flow sufficed for the needs of Tucsonans and their crops. In the past, this water was enough to power a grinding mill and to fill Silver Lake, which was Tucson's outdoor social center.

Left: Bend of the Santa Cruz, looking south from San Rafael Valley

A native Hispanic Tucsonan recalled those vanished times in words set down by Patricia Preciado Martin:

> They tore down the Otero house. They had a piano. The señora played beautifully. Everyone would come to listen. People would leave flowers on her doorstep. On Sundays we would all walk down for the *paseo* in the plaza of the Cathedral. Sometimes we would walk down to the river. In summers we would picnic under the cool shade of the cottonwoods. Everywhere there was music. Now the river is dry. The river had water then. The milpas are gone and the people are gone.

Now that the river has disappeared underground, ever-deepening wells are needed to supply the people and their fields, the orchards and mines that fill the river valley along a hundred-mile stretch. As the water-table goes down from ceaseless pumping, Tucson hopes that the Central Arizona Project will come to its aid before the century is out. This plan is to lift water two thousand feet from the Colorado River, three hundred miles away. Tucson's dream is doomed to end in nightmare if there should then not be enough water in the Colorado, or if the power to bring it should fail. It would not be the first time that the desert has called time on man who builds his city in its midst.

If a Tucsonan should recognize the presence and the importance of the Santa Cruz, it is unlikely that he would know where the river rises. "Somewhere down Mexico way," he would probably say. His lack of precise knowledge is understandable, for the river's origins are obscure. The River of the Holy Cross, named by Kino in the 1690s, rises in Arizona about fifty miles southeast of Tucson in the grasslands of the San Rafael Valley. It then flows south across the border into Sonora as far as San Lázaro, before a range of mountains turns the river and sends it back across the border five and a half miles east of Nogales.

As it continues north toward the Gila, the Santa Cruz gathers drainage from the Santa Rita, Tumacacori, Tucson, Rincon, and Santa Catalina ranges. Between Nogales and Tucson, Sonoita Creek was its chief tributary; its water is now impounded in Patagonia Lake.

Throughout millennia the Santa Cruz has greened its valley on each side of the border. Until people and power disrupted it, the valley was an idyllic place of plant, animal, and human life. Until a few generations ago, travellers were impressed by its lushness. Here is how it appeared in 1864 to J. Ross Browne:

> The valley of the Santa Cruz is one of the richest and most beautiful grazing and agricultural regions I have ever seen. Occasionally the river sinks, but even at these points the grass is abundant and luxurious. We travelled, league after league, through waving fields of grass, from two to four feet high, and this at a season when cattle were dying of starvation all over the middle and southern parts of California. Mesquite and cotton-wood are abundant, and there is no lack of water most of the way to Santa Cruz.

In flood the river rushed toward the Gila, spreading its silt over the plain, finally to lose most of its water in the desert before gaining the greater river. Today only the upper reaches of the Santa Cruz in Sonora indicate how the lower river too once looked.

Near Tucson it is now a drab sight. The bordering cottonwoods, alders, willows, and mesquites are gone, either cut for firewood or dead from lack of water. The riverbed has deepened from confinement. At the San Xavier bridge it is no longer possible for Papagos to irrigate their fields from surface takeoff. The Rillito is even lower, except when in flood it gathers the waters of its tributary washes, the Tanque Verde and the Pantano, and gives a costly lesson to those who settled on its flood plain.

Seen from the air, there is one fair point on the Santa Cruz's dry run through town — an unexpected oasis of pools and grass and trees. This is the plant where Tucson treats its sewage, then releases the purified effluent into the riverbed, where it is pumped off to irrigate fields of Pima cotton.

Beyond Marana and Picacho, the Santa Cruz widens into dwindling channels and becomes indistinguishable from the also lost Santa Rosa Wash, whose main channel is now blocked by the grandiose Tat Momolikot Dam.

This dam's ill-fated predecessor, the Santa Cruz Reservoir scheme, was another of the brain childs of Colonel William C. Greene, the copper king of Cananea. In 1908 he proposed to impound both the Santa Cruz and the Santa Rosa behind a dam near Picacho and also to divert the Gila to form another reservoir. Two million acre feet of water would ultimately reclaim four hundred thousand acres of desert land. Three hundred thousand dollars went into the construction of an earthen dam twenty-four feet high, twelve feet wide at the top, and a mile and a half long.

Greene's death in 1911 spared his seeing the dam swept away by flood in 1913, the fate of earlier dams on the rivers of Arizona since olden times. Today that vast acreage and more has been reclaimed, but not by dammed reservoir water. Pumps that are powered by electricity, whether generated by water power or steam plants, run day and night, inexorably lowering the subterranean reservoir. Such is the theme of Charles Bowden's sombre book, *Killing the Hidden Waters*.

A clear day in fall is a good time to reconnoitre the Santa Cruz upstream from the city. The freshly whitewashed towers of San Xavier provide an excuse for a detour to the mission, fondly called by Tucsonans the White Dove of the Desert. I once believed that its bare plaza should be graced by a fountain in memory of Francisco Garcés, the hardy Franciscan from Aragón who was the first priest of the present church. It was Garcés who pioneered the overland trail to Alta California for the Anza expedition of 1775–76. A friend who lives on the edge of the San Xavier reservation dissuaded me by pointing out that the Papagos, whom the mission serves, do not share our ideas of beauty. To them a well is more beautiful than a fountain.

Nowhere in Arizona is Father Garcés fittingly memorialized, although in Yuma near where he was martyred, there is a conventional statue by the Fulda brothers of him blessing the Indians. No place is named for him. One must go far to find the great limestone carving that ennobles a traffic circle in Bakersfield and commemorates the Franciscan's "discovery" of the lower San Joaquin Valley. Juan Bautista de Anza too is neglected. In all of Arizona no place bears his name. A Tucson drive-in movie theater is incorrectly called the De Anza.

Father Kino is well remembered in his Pimería Alta. There are statues of him in Phoenix and Hermosillo, Tucson and Nogales. In the latter town, a bronze by Ralph Hume seems to me best to honor this hardy Jesuit pioneer whose endurance has never been equalled.

Nogales is also graced by the most beautiful library building in the Southwest, an enchanting structure of native materials designed by Bennie Gonzalez.

That stretch of the river between Tucson and Nogales is an epitome of our history and culture. On the west side of the river, overlooking the retirement colony of Green Valley, are open-pit copper mines, made tolerable by the absence of smelters. Their ore is shipped by rail to the smelters at Douglas, San Manuel, or Hayden, to the atmospheric detriment of those communities.

To the east, on the Mt. Hopkins summit of the saddleback Santa Rita range, the world's first multiple mirror telescope scans the sky. Along the eastern shore of the Santa Cruz, a vast orchard of pecan trees takes a large volume of pumped groundwater. Heavy drinkers though they be, these nut trees, whether leafy or bare, make a fair border along a river that has lost most of its native silva. South of San Xavier, a mesquite forest has died from lack of surface water. It was once the summer haunt of white-wing doves.

Near the border, Tubac Presidio State Park and Tumacacori Mission National Monument commemorate the Spanish presence of crown and cross. At Tubac, Anza marshalled his colonists, their livestock, and supplies for the greatest overland trek in North American history. Tumacacori Mission was one of the three northernmost founded by Kino, the others being at Guevavi (now gone) and at Bac, where San Xavier still gleams to the glory of God. At *presidio* and mission, reverence for our past is paid. They are places of refreshment and inspiration for all who weary of frontier violence and the mock reenactment of shootings and hangings, all who believe that our hero priorities need reordering.

North from Nogales, refrigerated diesel rigs travel by day and night, carrying vegetables from the fields of Sonora and Sinaloa to the markets of North America. Back to the border roll the empties. Does the shade of Kino ever return to see the fruits of his husbandry?

Now the road leaves the river and crosses the Patagonias, a rugged range haunted by the ghosts of mining camps — Mowry, Harshaw, Washington — peaceful at last, man's lust for metals spent for the moment. From the summit, the view is over the grasslands of the San Rafael Valley. Here in another Shangri-la is where the Santa Cruz heads. To the north are the Canelo Hills, whose cinnamon color gives them their Spanish name; in the east lie the blue Huachucas, another of the desert-island ranges whose summits in winter yield snow-melt that forms springs and swells the tributaries of the Santa Cruz and the San Pedro. Summer rains green these grasslands; in winter they are sere. Until it was dammed to form a recreational lake in the Huachucas, Parker Creek was the Santa Cruz's main tributary in the valley.

On the border south of the Huachucas is the Coronado National Memorial, marking the conquistador's entry into what is now Arizona. The road to the monument — narrow, dusty, and beautiful beyond all freeways — wanders over the valley and threads the foothills, then crosses Montezuma Pass to enter the valley of the San Pedro.

Toward the end of the 1880s, the Cameron brothers, Colin and Brewster, came from Pennsylvania to run the first registered Herefords in Arizona on their six hundred thousand acre spread in this old land-grant San Rafael Valley. In 1900, Colin built a thirty-room French colonial mansion to serve as both residence and ranch headquarters. Situated on a dominant knoll, it was probably the grandest country house in all the Territory. In our time it has served as the ranch house in the movie version of the musical *Oklahoma,* although most of the film was made north of the Canelos near Elgin.

In 1909, Cameron sold the ranch to Colonel William C. Greene of Cananea, whose smelter still stands as a smoking landmark and source of fluvial pollution. Today, descendants of the Colonel still inhabit the great house. It is pictured in Janet Stewart's *Arizona Ranch Houses.*

The nearby border hamlet of La Noria was renamed Lochiel by the Camerons after their ancestral home in Scotland. Hispanos on both sides of the line still use the old name.

The 5000-feet elevation of the San Rafael Valley means rainfall and grass. Overgrazing and erosion have been checked. Trees are predominantly oak — Arizona, Emory, and Mexican Blue — with cottonwoods, sycamores, and willows along the nascent river and its creeks. Birds are Mearns' quail, uncrested Arizona bluejays, meadowlarks, and finches. Herons and cranes feed in the cienegas; the hawk patrol keeps down grain-eating rodents. Grazing Herefords pepper the slopes, and lone ranch pickups leave a dusty wake. Urban development has not touched this idyllic valley.

From the air, the headwaters of the Santa Cruz are hard to distinguish. On our flight up-river from Tucson, Chris took the Cessna higher and higher until at 14,000 feet, in bitter December air, we could only dimly perceive the river's course into Sonora and back into Arizona. The colors and contours of the land were subtle and gentle. The river's importance in history was apparent only to the imagination.

After talking by radio with an official at the Nogales airport, Chris brought the plane down at Sierra Vista, whose airport serves both the town and Fort Huachuca, a major Army electronics training-center. There we refueled and ate our sandwiches. Chris reported on his contact with the F. A. A.'s Flight Service Station.

Right: Along the Santa Cruz
Far right: Tucson

"I think I got through to him that we're not flying pot," he grinned. "I told him you're on a scientific mission, that you wanted to observe the river's Sonoran parabola. Right? You don't have any pot on you, do you?"

"Gave it up as a boy," I said.

"The trouble is," Chris continued, "I don't have Mexican entry papers in case we had to land."

"What would happen?"

"Plane would be scarfed and we'd go to jail."

"How about landing at Nogales and getting a permit?"

"I forgot to bring my ownership papers. No papers, no permit."

"I need to see that bend in the river from down low. Up where you took us, I couldn't make out a thing."

"My camera could," Michael said. "I got some beauties."

"How far is it into Sonora?" Chris asked.

"About forty miles."

"No way!"

"You don't *look* conservative," I said, eyeing Chris's beard, boots, and beat-up hat.

"Only when it comes to The Animal," he said, patting the plane.

I could only agree that the risk was too great, and so we limited the rest of the flight to the borderland between the San Pedro and the Santa Cruz, with a sidetrip in and out of Bisbee's Lavender Pit for Michael's benefit, then back to Tucson.

I knew that I had to see that bend in the Santa Cruz and also track the river to its source. And so one fine day in spring, after a winter of rainfall and flooding, four of us Tucsonans took the long way by land to the headwaters of the Santa Cruz. Buttressed by books and an ice chest of food and drink, we headed up-river from an early rendezvous at San Xavier. Between us we were formidably knowledgeable in history, archaeology, dendrochronology, bibliothechnics, linguistics, and literature; and yet we took the wrong turn heading south from Nogales and dead-ended at an army barracks, sentries at carbines ready.

Our *¿Donde esta el camino a San Lázaro?* brought an answer that got us on a dusty road that led to the river. Along its west bank we proceeded at a leisurely pace, each volunteering his specialized knowledge of the landscape. In the stretch between Nogales and San Lázaro, the road climbs a gradual five hundred feet, and as much again before it reaches the river's source in Arizona.

The oak and mesquite woodland appeared sorely damaged by January's cold. Towering cottonwoods recalled to mind the way the land looked before downstream Arizonans axed and burned out the thirsty trees. At San Lázaro I saw at last what I wanted to see: the bend in the Santa Cruz that turns it back toward Arizona. It is a far shorter curve than the great bend on the Gila.

From a rise above the village, we gazed on the narrow canyon that carries the river through the foothills of the Sierra de San Antonio, a continuation of the Patagonias. We also saw a community which, except for power acquired by agreement with the U.S. Rural Electrification Administration, is essentially unchanged since its beginnings. Here is how all of the Pimería Alta once looked. We had come far back in time.

Our portable library was now called upon for the accounts of Bartlett (1851) and Browne (1864), who came this way and testified to the toll exacted by the Apaches from those who dared settle athwart their Sonora raiding trail. We followed the river along its northerly course, finding all quiet in the larger village of Santa Cruz. At a point on the river recognizable from the drawing by Ross Browne, we opened the ice chest and ate our lunch in front of an audience of steers and burros.

The border crossing at Lochiel-La Noria probably carries the least traffic on all the United States-Mexico frontier. We were reluctant to disturb the somnolent Mexican official in his rocking chair on the porch. In accepting our document of passage with barely a glance, he muttered *"Bueno"* and resumed his nap.

From there we headed across the San Rafael Valley to the old Cameron house. Although confined in a narrow channel, the Santa Cruz was alive with cool water and bordered by cottonwoods and willows. Was that a blue heron or a sandhill crane that rose to fly a safe distance downstream, lower its legs, and resume feeding? Across the road ahead, four pronghorn antelopes were all but airborne, then gave us their white rumps as they turned with pricked ears and froze for pictures.

Now as the grassland began to roll and rise toward the low range of the Canelo Hills, the young river lost its identity. We were unsure which was the main fork, if any. Day was closing down and we still had miles to go to reach the city. And so we reluctantly left it to another day to determine precisely where the Santa Cruz heads.

It was early in October when, after a long, hot summer, a friend and I set out again in the Scout to find the veritable sources of the river. Bolstered by the Forest Service map of the lower Coronado National Forest and by information from friends at the Geological Survey and the University of Arizona's Paleoenvironmental Laboratory, we headed up Patagonia Creek toward the turnoff to Saddle Mountain and Meadow Valley Flat.

We found it to be a dirt track just over the divide between the creek and San Rafael Valley. Soon we came within sight of one of our goals, a waterfowl nesting impoundment created in the early 1970s by the Arizona Department of Game and Fish at the Bog Hole, a cienega on Meadow Valley Creek.

We proceeded on foot over short-cropped grassland, surmounted a barbed wire enclosure fence that surrounds the sanctuary to prevent grazing, and reached a low earthen dam forming a pond of several acres. The plan was to create similar impoundments on neighboring creeks where cienegas or, in cruder Anglo-Saxon, bog holes occur. The beneficiary was to be the endangered Mexican duck, whose migratory flyway is over the San Rafael. Opposition by cattlemen to the loss of grazing land, and also by environmentalists who object to any tampering with topography, has thus far limited the project to the single pond. Also prevented by protest was an ill-advised project to establish a prairie dog colony in the area.

Water level is maintained in the Bog Hole pond from seep and seasonal runoff. Below the dam, intermittent flow supports frogs enough to feed many a heron.

So we had come at last to one of the sources of the Santa Cruz. We found another source a mile east at some seeping springs in Sheep Ranch Canyon, above the headquarters of the Vaca Ranch. There we talked with Foreman Hudson and his son — laconic, weathered, working cattlemen. They assured us that not far down-creek was the junction of Bog Hole and Sheep Canyon creeks, which together with Cherry Creek off the Huachucas and Mowry Wash down from the Patagonias give the Santa Cruz its initial identity.

And so it was. We parked at an overlook where grassland, juniper, Emory oak, cottonwood, and willow grow. There we broke out the ice chest. As we lunched, our gaze was over an *arroyo seco* where, earlier in the year, we had seen a little river beginning life under its own name, El Río de la Santa Cruz.

All that was needed to restore the river to life, from its sources to its end short of the Gila, was rain — rain on the Canelos and neighboring ranges that would swell seep and trickle to brook and flood and thereby reactivate the cyclic flow, one as beneficent to Tucson as that of the Salt is to Phoenix.

Río Gila

AS THE RIO GRANDE DIVIDES NEW MEXICO ALONG ITS NORTH-SOUTH course, so does the Río Gila divide Arizona on its way from east to west. Whereas New Mexico is left essentially intact by the division, Arizona is separated culturally as well as geographically by this river that crosses the state from New Mexico to California.

Up to the Gila extended the Pimería Alta, the land of the Upper Pimas. There at the river, the Indo-Spanish-Mexican influence ended. In 1848 the Treaty of Guadalupe Hidalgo established the Gila as the boundary between Mexico and the United States. The Gadsden Purchase five years later gave the United States the land south of the Gila to the present-day border. Yet to this day the river remains a natural boundary.

North of the Gila rise the terminal mountains of the Colorado Plateau; south of the Gila, the Sonoran desert extends deep into Mexico. North of the river are the Anglo cities of Prescott and Phoenix; south of the river, Sonoran Tucson and Yuma. Regardless of the political boundary, north of the Gila is Anglo-American, south of the Gila is Indo-Sonoran. These contrasts can be multiplied to the degree that Arizona appears as two states.

Left: Flooding on the Gila above Painted Rock Dam

Before dams, canals, and pumping made its flow intermittent rather than perennial, the Gila was the lower Colorado's main tributary. No longer does it carry abundant water and silt. Along much of its length, the Gila is a dead river. A recent prize essay by historian Henry F. Dobyns is called "Who Killed the Gila?"

His answer is the Anglos who came after the United States seized the land from Mexico in 1846 and the Civil War gave Arizona territorial status in 1863. Whereas the Mexicans had never colonized the Gila's valleys and were not affected by the Anglos' pre-emption of the water, the culture of the riverine Pimas, descendants of the Hohokam, was brought to a dry end.

Here is a lament by George Webb, a Pima elder, whose pastoral book *A Pima Remembers* is an elegiac classic:

In the old days, on hot summer nights, a low mist would spread over the river and the sloughs. Then the sun would come up and the mist would disappear. On those hot nights the cattle often gathered along the river up to their knees in the cool mud.

Soon some Pima boy would come along and dive into the big ditch and swim for awhile. Then he would get out and open the headgate and the water would come splashing into the laterals and flow out along the ditches. By this time all the Pimas were out in the fields with their shovels. They would fan out and lead the water to the alfalfa, along the corn rows, and over to the melons. The red-wing blackbirds would sing in the trees and fly down to look for bugs along the ditches. Their song always means that there is water close by as they will not sing if there is not water splashing somewhere.

The green of those Pima fields spread along the river for many miles in the old days when there was plenty of water. Now the river is an empty bed full of sand. Now you can stand in that same place and see the wind tearing pieces of bark off the cottonwood trees along the dry ditches. The dead trees stand there like white bones. The red-wing blackbirds have gone somewhere else. Mesquite and brush and tumbleweeds have begun to turn those Pima fields back into desert.

Now you can look out across the valley and see the green alfalfa and cotton spreading for miles on the farms of white people who irrigate their land with hundreds of pumps running night and day. Some of those farms take their water from big ditches dug hundreds of years ago by Pimas, or the ancestors of Pimas. Over there across the valley is where the red-wing blackbirds are singing today.

How did the Gila come by its name? The answer depends upon uncertain sources. Although Fray Marcos and Vásquez de Coronado crossed the Gila en route to Zuñi, they left no reference to the river. When Alarcón sailed up the Colorado in 1540 to the mouth of the Gila, he named the tributary the Miraflores.

In a translation of Benavides's *Memorial* of 1630 appears a foot-note by Hodge that reads: "Benavides is the first person to use the name which today survives in the Rio 'Gila' of Arizona and New Mexico. At the period of our author, Xila was only the name of an Apache settlement. . . in Socorro, New Mexico." An entry in *Arizona Place Names* casts more light: "In 1630 a province of New Mexico was named Xila, or Gila, a Spanish word encountered on maps of Spain itself and used in the language as an idiomatic expression: *de Gila,* 'a steady going to or from a place.' "

How the name and the river became linked remains a mystery. We only know that it first occurred on Kino's 1695 map of the Pimería Alta. There we read *Río de Hila.* Kino earlier called it the River of the Apostles and its tributaries — Santa Cruz, San Pedro, Verde, and Salado or Salt — the Four Evangels.

Throughout these permutations the name has persisted, however spelled and whatever its pronunciation: Jeela, Guyla, or Helay, the last being, according to James Pattie, the one used by the trappers who came in the 1820s and trapped out the beavers along the Gila and its tributaries. This hastened the river's demise as a perennial stream by disrupting the ecological balance.

Although the Gila is predominantly an Arizona river, it rises in New Mexico. Up near its sources in the Mogollon wilderness along the Continental Divide, the Gila is still a beautiful living river.

I went to see for myself, starting out at Silver City in the heart of the copper country. Here is the erstwhile domain of the Mimbres Apaches, whose great chief was Mangas Coloradas — Red Sleeves. East from the Divide, the Mimbres River flows to the Río Grande, while on the westward slope, the three forks of the Gila meet near the present Gila Cliff Dwellers National Monument. I came after a winter of record flooding in which bridges were swept away and communication lines downed for weeks. The Gila's bed was still a tangle of uprooted cottonwoods. It was easy to understand how, in the years before dams brought some flood control, the river could gather such volume in its six hundred mile course as to make it second only to the Colorado in powerful flow.

Guided by Kit Carson, General Stephen W. Kearny and his Army of the Pacific, composed of a small troop of dragoons with pack mules and brass howitzers, came this way in the fall of 1846 on their forced march to add California to their bloodless conquest of New Mexico. Up the Mimbres and down the Gila they toiled through a terrain of gorges and cliffs, forest and underbrush. To avoid this hazardous route, Lieutenant Philip St. George Cooke and his battalion of Mormon volunteers and their wagons took a more southerly way around the mountains, over the San Pedro and down the Santa Cruz to meet the Gila near the Pima villages and thence follow in Kearny's tracks to the coast.

Kearny's adjutant, Captain Henry Turner, and his topographical engineer, Major William Emory, kept journals on the trek. On October 20, 1846, Turner reported reaching the upper Gila: "A beautiful stream — perfectly clear water, and about thirty steps across, timbered with cottonwood principally." Then he recorded days of broken marches when the miniature army was forced to leave the canyoned river and seek passage

> . . . up and down steep mountains, over points and ravines jagged with sharp rocks, making it most laborious, and almost dangerous for our mules. At last we came to the bluff overlooking the river, the sides of which seemed more precipitous, if possible, than our ascent, but we descended safely five or six miles further, and are again encamped on the Gila, where as yesterday, we caught an abundance of fine fish, and find the grass tolerably good. In consequence of the bad road today, the howitzers have not come up. Partridges and turkeys are abundant on the bottom of the river, also deer and bear and beaver in great quantity.

Captain Turner was right in predicting that those kanyons [sic] of the Gila would never be the trail to California. They lie in some of the roughest country in the West. In 1924, thanks to the missionary leadership of Aldo Leopold, the canyons were given inviolable status as the first National Wilderness area, access to which is only by foot or horse. In recent years, when owners of a four-wheel-drive vehicle defied the ban and drove up near the source of the Gila's main fork, they were required to dismantle and pack it out in parts.

When Cooke and his Mormons reached the middle Gila, between the junctions of the San Pedro and the Santa Cruz, he found the Pimas ready to supply the soldiers with melons, squash, beans, and mules.

Cooke had an eye for beauty, as we learn from this entry in his journal: "I rode up to a group of girls naked to the hips. It was a gladdening sight. One little girl excited much interest with me. She was so joyous that she seemed very innocent and pretty. I could not resist tying a red silk handkerchief on her head for a turban."

Turner also had favorable things to say of this tribe:

> These Pimas are a good harmless people in appearance, and more industrious than I have ever found Indians — they have all the necessities of life in sufficient abundance, and all produced by their own industry. . . . The men generally have kind, amiable expressions — never did I look upon a more benevolent face than that of the old chief — he is a man of about sixty years of age — spare and tall, and exhibits more of human kindness in his face, air and manner than I have ever seen in any other single individual.

Today these Indians form the Gila River tribe. The tribal seat of their reservation is Sacaton. Few signs of prosperity other than pickup trucks and mobile homes are to be seen. Few of the necessities of life are produced locally. Cooke would find the town a disheartening sight, and Turner would be dismayed by the decline of this once thriving people. Pima Elder Webb's lament (see page 21) is all too pertinent. What water there is comes from ever-deepening sources in the ground reservoir.

Up-river from Sacaton is the mysterious Casa Grande. It has been recorded in words and drawings by travellers from the time of Kino's *entrada* of 1694, when he gave it the name it still bears. Three years later, Kino and his party took shelter there in a November storm, celebrated mass, and broke fast. In his *Favores Celestiales* Kino left this record:

> The soldiers were much delighted to see the Casa Grande. We marveled at seeing that it was about a league from the river and without water; but afterward we saw that it had a large aqueduct with a very great embankment, which must have been three varas high and six or seven wide — wider than the causeway of Guadalupe at Mexico. This very great aqueduct, as is still seen, not only conducted the water from the river to the Casa Grande, but at the same time, making a great turn, it watered and enclosed a plain many leagues in length and breadth and of very level and very rich land. With ease, also, one could now restore and roof the house and repair the great aqueduct for a very good pueblo, for near by there are six or seven villages of Pimas Sobaipuris.

Although the purpose of the Casa Grande is unknown, we do know that a kinfolk of the Hohokam, called the Salado, reared the four-story structure of rammed earth around A.D. 1350. They occupied it for only a century and then they too went away. Today it is administered by the Park Service as a national monument.

Down-river from Casa Grande, beyond the crossing of Interstate 10 at Gila Butte, is the site of a 1400-acre Hohokam community. When Pimas resettled the place in the late 1870s, they called it Snaketown. Its excavation and description are among Emil Haury's most distinguished achievements in Southwestern archaeology. Here were found traces of the oldest irrigation system in the United States. Snaketown has now come under the jurisdiction of the Park Service as the Hohokam-Pima National Monument. Although the Hohokam's tenancy of this site endured for nearly two thousand years, it too came to an end, one more evidence of man as leaseholder, not owner, of the land.

Until masonry and concrete were used to build dams, man proved powerless to arrest the Gila in flood. Brush- and rock-filled weirs were intended by Indians and Anglos only for diversion purposes. When the river reached flood stage, such barriers were swept away like straws, even as the one built by Colonel Greene on the Santa Cruz. Beginning in 1900, repeated attempts were made to dam the Gila below the mouth of the Hassayampa River. Wood and rock, earth and brush all proved ineffectual. Success did not come until 1921 when Frank Gillespie, a prosperous cotton farmer and water user, built a concrete dam here, which stands to this day. Its reservoir also impounds the flood waters of Centennial Wash. Below the dam, nearing the big bend of the Gila, lie thousands of acres of Arizona's richest cotton lands.

In 1959, Painted Rock Dam was built down-river from Gila Bend, near the ancient petroglyphic formation sketched by Ross Browne and which is now a state historic park. This dam affords flood protection and irrigation water for the citrus and cotton plantings that green the valley from Gila Bend to Yuma.

Right: Coolidge Dam on the Gila River
Far right: The great bend of the Gila

Another rich farming area was developed after the mid nineteenth century by Americans, especially Mormons, along the middle reaches of the Gila up-river from Florence to where the stream breaks through the mountains and gathers the waters of the San Pedro. A main diversion canal now carries the flow past the town and the state prison and the neighboring town of Coolidge to the Picacho reservoir, where it is drawn on to irrigate cotton, alfalfa, wheat, and barley.

It was this major takeoff that dried up the lower reaches and led to George Webb's lament quoted earlier (page 21). When the Gila's downstream sources of water from the Salt and the Verde rivers were pre-empted by Phoenix and the Salt River Water Users' Association, the Gila again ran dry until it received increments from the New, Agua Fria, and Hassayampa rivers.

In the late 1920s the Bureau of Reclamation and the Army Corps of Engineers dammed the Gila below the mouth of the San Carlos River at the point where the Gila enters a narrow canyon between the Pinal and Mescal ranges. This is the heart of the San Carlos Apache country, given immortality in the works of Ross Santee, Arizona's horse wrangler-writer-artist. His ashes were scattered on the slopes of the Pinals where the sotol blooms.

Coolidge Dam formed San Carlos Lake, which inundated what the Apaches held to be one of their ancient burial grounds. Controversy was inevitable, and it continued over who had jurisdiction on fishing rights in the lake — the tribe or the Arizona Department of Game and Fish. This was finally resolved in favor of the Apaches.

The lake was slow to fill during years of subnormal rainfall. Not until a half century after the dam was dedicated by former President Calvin Coolidge at a ceremony enlivened by Will Rogers's quip, "Mr. President, if this was my lake, I'd mow it," — not until 1979, after two years of unprecedented rainfall in the mountains of New Mexico and Arizona, was Lake San Carlos filled to capacity. Then the dam was forced to release water in such amounts as to bring the Gila to life.

When I came down-river from the Gila wilderness, Lake San Carlos was backed up halfway to Safford. Disastrous flooding and erosion had occurred above and below that Mormon town. Not until another dam is built on the Gila, probably below its junction with the San Francisco — which also gathers the water of the Blue — will the Safford area be safe from recurrent disasters.

Safford's guardian peak is Mt. Graham. At 10,753 foot elevation, it is the third highest point in Arizona after the San Francisco Peaks and Mt. Baldy. This great blue-shouldered mountain stands over the Gila Valley just as Mt. San Jacinto rises above the Coachella Valley in California.

Since my first sight of Graham from a viewpoint in the Santa Catalinas, I had seen the mountain from all sides yet never ascended it. Now on my way down-river I took advantage of a newly surfaced road to drive nearly to the top. From there, mine was a reverse view of the one of twenty-five years ago, west a hundred miles over the Sulphur Spring Valley, the Galiuros, and the San Pedro to the cloud that was Mt. Lemmon. To the east beyond the San Simon Valley was New Mexico and the end of the Mogollon Rim. Below in the west lay the route of Fray Marcos and Coronado. Mt. Graham is heavily forested and still the haunt of black bears. From its springs and lakes, water makes its way down the north slopes to the Gila. Here is another of Arizona's desert islands in the sky.

Down-river the next day I passed the Apache settlement of Bylas near where it is believed that the Coronado expedition emerged from the pass between Mt. Graham and Mt. Turnbull, forded the river, and entered the mountains that led to Zuñi. Along this hundred-mile stretch of the river, one passes from Mormon country along the edge of the San Carlos Apache domain to the American copper mines at Globe-Miami.

The road along the southern side of the lake led to Coolidge Dam. There the water was up to the lip of the spillway; the cowboy humorist's quip of fifty years ago was no longer pertinent. The lake had since risen into every least arroyo that had hitherto known only scant rainfall.

In that rainy winter, the dams on the Gila released their water and the river ran again along its entire length. Again I took to the air with Chris and Michael to see what the flooding looked like from above.

Flying out of Tucson we climbed high for a view over the big bend. The weather was clear and very cold, and I was glad to be wearing a down-filled jacket. My young colleagues impressed me by their obliviousness to the weather, dressed as they were in jeans and cotton flannel shirts, until I learned that the plane's heater was up front between them. In spite of the bitter cold air from the open window, through which Michael was shooting, the gently rocking wings lulled me into a sense of perfect security.

Between Gila Bend and Painted Rock Dam, the backed-up river had drowned fields and farms to a depth that revealed only the topmost fronds of tall palms. If the damage to property were overlooked, the disaster could be seen as a boon to agriculture. The aquifer was recharged, the canals below Painted Rock were silvery full, and the land had gained a fresh deposit of silt.

We flew on to Yuma and landed at the Marine Base to refuel. Small private planes have a low priority, as military and commercial aircraft also land and take off there in what seemed a bewildering flight pattern. Condit was at home, however, as he had once been based at Yuma while flying for the U.S. Geological Survey.

Heading home, we followed the Gila to the bend, descending nearly to ground level for Michael to shoot a fleet of big green cotton-picking machines. We came low enough to exchange waves with the drivers — or were they shaking their fists?

In the late afternoon the air was clear and smooth. At 140 m.p.h. we were cradled on tranquil air. Visibility was over nine separate ranges to the plumed smelter at Ajo, and in the other direction to the snow-capped San Francisco Peaks.

Unless they are running high and overflowing, Arizona's rivers are not much to see from above, and yet in whatever season, they are the essential resources of a land that would otherwise be as barren as the moon. As a reading traveller and a travelling reader, I have come to see the land with the eyes of all who have gone before. This familiarity lends security whether on land or in the air, oriented as I am without reference to maps and ever respectful toward those who travelled without benefit of our topographical knowledge.

As we crossed Santa Rosa Wash, Chris circled for a better view of another dam. This was the Tat Momolikot of dubious value, built across the wash by the Army Engineers. Its purpose was to protect the downstream cotton plantings of agribusiness and also to provide the Papagos on whose land the dam lay with a lucrative resort lake.

Alas for the latter purpose. In the years since the four-mile-wide barrier was constructed, no lake has ever been formed. Even after the recent rainy seasons, barely a puddle was there. Now the Papagos can add "boondoggle" to their vocabulary.

Day was ending as the lights of Tucson came up ahead and we were talked down by rapid-fire instructions from the tower. In we silently glided, engine throttled down, earthlings once again.

Río Salado

KINO NAMED IT IN 1698 THE RIO SALADO FROM A STRETCH OF saline beds over which it flowed. It was then a perennial stream. Now by the time the river has reached Phoenix, its water has been diverted. As it straggles through the city to reach the Gila, the Salt's course is a sad sight, good only for sand and gravel. The river's life has been given to the Valley of the Sun and the risen Phoenix, now the agricultural, industrial, and urban heart of Arizona. There in one of the driest deserts is one of the greenest oases.

Water has done it, river water from the mountain backbone of the state — the Mogollon plateau that draws down the rain and the snow. Throughout the millennia of this cyclic flow, soil was also brought down by the river and spread over the valley floor. Add sun to soil and water, and growth was inevitable, of seed and leaf, flower and fruit. And a great city.

The lack of rainfall that would otherwise have left the area an unreclaimed desert was compensated for by the perennial flow of river water, to which the Salt's major tributary, the Verde, contributes a large share. This plentitude has made Phoenix the power center of the state. Tuc-son's Río de la Santa Cruz has no Mogollon Rim to swell it; and as we have seen, its meagre flow has been further diminished by mines, orchards, and the city, until it now traverses Tucson only in floods that are gone in a flash.

Although Phoenix is the first city to rise there — its explosive growth came after World War II — it is not the first civilization to appear in the sun-drenched valley of the Salt. The Hohokam were skilled water engineers. Although unable to arrest the river for flood control and irrigation, they were able to dig major canals to carry water to their fields and dwellings, sited far enough from the river to be safe from flooding.

In 1887–88 the Southwest's first archaeological exploration was carried out at a Hohokam site six miles from the river. There the Mary Hemenway expedition, led by Frank Hamilton Cushing and with the young Frederick Webb Hodge as secretary, unearthed a buried community and sent the artifacts to the Peabody Museum at Harvard. There they languished until the early 1930s, when they became the subject of Emil Haury's dissertation for his doctorate, published in 1945. He called the settlement La Ciudad de los Muertos from the number of skeletons that were found. Today the site has disappeared again, this time beneath the spreading city of Tempe.

Left: Confluence of the White and Black rivers

Why the Hohokam went away or were all used up, no one knows for sure. Was it a drought such as the one that lasted thirty years and ended the culture of the cliff dwellers of northern Arizona? In such a period the Salt would have run dry. Was it over-irrigation that salinized the land? Or did raiding tribes drive away the river farmers? Haury believes that cultures throughout the Southwest, not only the riverine Hohokam, came to a cyclic end by A.D. 1450.

What the Hohokam left were sand-filled canals and vestigial dwellings. When increasing numbers of Anglos arrived in the valley after Arizona became a territory and troops were posted at Fort McDowell near the confluence of the Salt and Verde, they soon discovered the valley's fertility. Horses needed hay and grain; men lived on meat, bread, potatoes, and fruit. And so farming began again along the Salt River, although it proved precarious because of flooding.

In 1867 an adventurer named Jack Swilling was engaged in hauling hay to Fort McDowell, a hazardous occupation in which the two previous haulers had been killed by Tonto Apaches. When Swilling observed traces of the canals of the Hohokam, he had a fertile idea. Digging was safer than hauling. Together with John "Yours Truly" Smith, Henry Wickenburg, and Louis J. F. Jaeger, he formed the Swilling Irrigating Canal Company. Near today's downtown Phoenix, and following a Hohokam ditch, he and sixteen cohorts dug the first modern canal, and that mythical bird began to rise from its sandy ruins. On the grounds of the present Greyhound Park, near Hohokam vestiges preserved as the Pueblo Grande, Swilling built an adobe hacienda for his Mexican bride, Trinidad. His time in the sun was brief. Alcohol and opium were his downfall, and he died in the Yuma prison in 1878, charged with highway robbery.

The townsite of Phoenix was incorporated in 1868. It was named by another adventurer, an English remittance man, Brian Philip Darrell Duppa, a self-styled Lord from Holsingbourne House, County Kent.

One day he and some convivial cronies were lounging at the Pueblo Grande, trying to settle on what to call the new town. It needed a name so that shippers could be instructed where to send supplies. Among those suggested was that of Pumpkinville. That was too much for "Lord" Duppa. He leapt on the wall and proclaimed, "As the mythical bird rose from its ashes on the Arabian Desert, so shall this town rise from the ruins of an earlier civilization. I name thee Phoenix."

As settlers came in mounting numbers, modern Arizona was born. The Salt River Valley Water Users' Association was formed. Whereas canals were easy to dig and extend, often following those of the Hohokam, flood control was difficult. Brush weirs and rock barriers were swept away when the spring thaw sent the Salt and the Verde roaring down on the valley.

Then a major turn in western history came in 1902, when President Theodore Roosevelt signed the Reclamation Act. Its author, Major John Wesley Powell, lay dying when word was brought of the signing. He was content. His had been a long struggle to convince the wet East that in the dry West he who owned the water owned all.

The harnessing of the Salt River began with Theodore Roosevelt Dam, completed in 1911. Its construction may be seen as the greatest single act in Arizona's history. Much of the state's development flowed from this structure and its stored water. President Roosevelt came for the dedication, and the ringing words he spoke still echo:

> I do not know if it is of any consequence to a man whether he has a statue after he is dead. If there could be any monument that would appeal to a man, surely it is this. You could not have done anything that would have pleased and touched me more than to name this great dam after me, and I thank you from my heart for having done so.

The rest is the history of the valley and its cities — Phoenix, Tempe,

Scottsdale, Mesa, and their satellite communities — as well as of the Salt River Project, which supplies water and power to a wide area. Below the Roosevelt is now a chain of dams and lakes — Granite Reef, Stewart Mountain, Mormon Flat, and Horse Mesa on the Salt; Bartlett and Horseshoe on the Verde — all forming a complex of rivers to be controlled for irrigation, electrical power, and recreation. Throughout the valley, canals distribute the precious element to homes and gardens, fields and orchards, and to the industries that have been drawn by water and sun to locate in greater Phoenix.

The bed of the Salt River that passes through the southern edge of Phoenix is not always dry. When heavy rainfall on the watersheds of the Salt and the Verde fills their dams to capacity, as in 1978, 1979, and 1980, the floodgates are opened and the river rampages through to the Gila. Then the city suffers from washed-out roads and bridges. Sky Harbor Airport is partly built on the old flood plain and its extended runways are sometimes under water. Phoenix will not be secure from these periodic disasters until it does what greater Los Angeles did with its once wild rivers and confines the Salt within concrete banks.

West of Phoenix on the road to Gila Bend, the land is neither town nor country. Feed lots, gravel pits, dumps now called Sanitary Refuse Disposal Facilities, trailer parks, and massage parlors sprawl in unplanned ugliness.

Near where the Gila enters the valley through a mountain gap, it meets the Salt, to the whine, roar, and smell of hot machinery from the Phoenix Auto Raceway. There in a salt cedar and tamarisk tangle of seep and stink, these once proud rivers join in what is the end of one and the continuation of the other. There is intermittent flow along the Gila's remaining course, past Gillespie Dam and around the great bend, then down valley past Painted Rock to the Colorado.

Not far southeast from the Salt and Gila's junction is the point on Monument Hill from which nearly all of Arizona's land measurements begin. It is the Gila and Salt River Base and Meridian point — where the Base Line and Meridian intersect, abbreviated on maps and deeds as the "G and SR BM."

There was nothing to keep me there once I saw that the Salt had met its sorry end. A backtrack led through Phoenix, Tempe, and Mesa to Apache Junction and the turnoff through the western reaches of the Superstitions to rejoin the Salt along its chain of dams and lakes. Access to the lakes is open to boating and camping; the dams are behind locked gates, all but the Roosevelt.

Time and again I have drawn up at this greatest of masonry dams, faced with reddish-brown sandstone quarried from the river canyon's walls. The lake it impounds, formed by the Salt River and Tonto Creek, extends deep into the Tonto Basin. The road northwest leads to Punkin Center, Sunflower, Payson, and the Rim. Zane Grey's Arizona home was near there. His cabin may still be seen. The road to the east heads roundabout for the twin copper towns of Globe and Miami.

Globe is the seat of Gila County, renowned for more than mines and smelters and the Apaches of the adjoining San Carlos reservation. Globe was the long-time stronghold of George Wiley Paul Hunt, the first and seven-times governor of Arizona. Up the steep streets of Globe he trudged as a youth, leading a pack burro laden with his worldly goods. There he waited table and tended store, toiled and rose to become the strong man of his time.

Ross Santee is also from Globe. The books he wrote and illustrated are centered in this mountain town and surrounding country. There

Following pages: Theodore Roosevelt Dam on the Salt River; Salt banks on the Salt River

are two Arizonas, one to be seen in *Arizona Highways*'s flaming skies and colored cliffs. The other has only the purity of light and the immensity of distance. Ross Santee is the laureate of this austere Arizona.

I passed through Globe and regained the Salt at the crossing of its deepest canyon. Down-river from here are the salt banks that led Kino to give the river its name. The road switchbacks down and then up to the forested Rim and the Mormon towns of Show Low, St. Johns, and Springerville, and the headwaters of the Little Colorado.

Once more I backtracked, this time in search of a road that would lead to the forks, the Black and the White, that form the Salt. Although reservation roads appear on the map, they are not always posted where they turn off the main highways. On what promised to be Road 4, I headed east into piñon and juniper lands. The dirt road wandered through a clean and lonely world. A sign proclaimed it to be Indian land, open to fishing and hunting only by permit from tribal headquarters at San Carlos.

I wanted only to cruise through, trusting that I would reach the Black and then the White. Unmarked divergences led me to pause and then continue. Small conifers yielded to tall ponderosas. Lumbering country began as the road climbed into the White Mountains.

Then I came to a deserted sawmill. The map showed that from there I could go east-southeast and gain the mountains above the Gila and again cross the route of Coronado. Instead I turned north on Road 9, which led to the river's forks. The road was graded but narrow, with occasional turnouts. I hoped that the idle sawmill meant no logging trucks. To meet a hell-bent Apache logger while trespassing on Apache land was not a good prospect, for nothing could stop one of those rigs on the downgrade.

Meet one I did, luckily at a turnout. He went by with a blast of air-horn in a smother of dust. Not far behind came a pickup that drew alongside and stopped. Its door bore the insignia of the San Carlos

Apache tribe. The driver was a uniformed young Apache wearing badge and gun. His unsmiling face looked me over.

"Need help?"

"I might have if I'd met that truck any other place."

"You wouldn't have needed any!"

"How far is it to the Black?"

"Ten miles or so. May I see your permit?"

"I'm not fishing, just driving through."

"O. K. Then you don't need one."

"Are you patrolling?"

"On my way back to San Carlos. One of our boys got locked up last night in Whiteriver." He was referring to the neighboring Apaches on the Fort Apache Reservation.

"Where does their land start?"

"At the Black. Be sure you close the gates. They've got more cattle than we have." He grinned for the first time. "They've got more of everything."

"You've got the horses," I said. "I passed a *remuda* back a few miles. That's where I found this." I pointed to a rusty horseshoe that I'd hung on the dashboard. "Who wrangles them?"

"Cowboys. Apache cowboys." He grinned again. "We can still ride."

"You've been to school," I said. "Where?"

"A.S.U. I plan to go to N.A.U. for forest management."

"You've got the trees!"

At the bottom of a steep descent I stopped on the high bridge over the Black River and looked down. Here was the main fork of the Salt, flowing fast over dark rocks, a big stream even in autumn. The river

was lined with great white-trunked sycamores; a bald eagle was perched on the highest tree. Clouds were gathering in the west — a weather front was approaching.

I was still far from the headwaters of the Black on the eastern slopes of Mt. Baldy (11,590), near the New Mexico line. Along its wild course the river gathers water from a hundred creeks, their names a litany of geographical poetry: Willow, Burnt Corral, and Bonita; Hurricane, Turkey, and Indian. No road goes there, and I would never make it on foot. That the Black rose and ran and was joined by the White to form the Salt and bring life and light to the lowlands, would have to be enough for me.

Across the bridge lay the Fort Apache Reservation. The San Carlos policeman was right — this was cattle country. I went through gate after gate — Texas gates of sapling stakes and barbed wire, easier to open than to close, that needed all my strength. I met no more cars.

The White is the lesser fork, rising on the northern slopes of Baldy, flowing a shorter distance and bringing less water to the Salt. It, too, has its creeks, including the Trout, Diamond, and Rock. Soon after crossing the White, I gained a highway and rolled on into Whiteriver, headquarters of the other Apache reservation in Arizona. I had been there before in 1960, trailing Martha Summerhayes and her army husband. At Fort Apache in 1875 she had borne her first child under primitive conditions. Earlier, in witnessing a Devil Dance, she had been enthralled by the virile Chief Diablo. The old fort buildings are still there, now serving as an Indian school.

Whiteriver is a rough town. There is a lot of life there, but not for an Anglo dude. I drove on through, heading for Show Low, a log fire, and bed for the night. Again my luck held. The weather front arrived in the night, and I woke to a snowy world. Yesterday's forest roads would have been unplowed and impassable even for the Scout.

On my partial exploration of the Salt River, I had come a far piece from its Gila junction. It is a journey I recommend to those Phoenicians who may be curious as to the source of their life in that oasis.

Río Verde

ALTHOUGH IT DOES NOT RIVAL THE COLORADO IN FLOW NOR THE Gila in length, the Río Verde has enough archaeological and topographical interest to set it apart from the other rivers of Arizona. Like the Santa Cruz, the Verde comes to a head in high grassland. Then, unlike the southern stream, it gathers water from creeks off the Rim and for the rest of its run to the reservoirs at Horseshoe and Bartlett dams is a perennial river. As the Salt's chief tributary, the Verde is an essential part of the Salt River Project, which supplies Phoenix and the Valley of the Sun with agricultural and domestic water and electrical power.

Along its middle reaches at what is the geographical center of Arizona, the Verde traverses a someways idyllic heartland valley. Here cottonwoods still flourish and copper-rich Mingus Mountain walls the west, while to the northeast the red rocks of Sedona shame even the chromatic pages of *Arizona Highways*. In this pastoral valley and on Beaver Creek are the restored ruins of Tuzigoot and Montezuma Castle, monuments in the National Parks system. Down-river where the Verde cuts a canyon through the mountains on its way to the Salt, Arizona's first hydroelectric plant, built in 1909, is still generating power.

Left: Sheep bridge on the Verde above Horseshoe Lake

To one seeing the middle Verde in summer when bordering cottonwoods hide the river, it is obvious why it is called the Green. Who first named it Verde? As we have seen, most roads in Arizona nomenclature lead back to Father Kino, that tireless traveller and namer. Writing in 1748 of Kino's northern *entrada* of 1699, Father Sedlmayr stated, "We have it from Captain Manje [Kino's military escort] that the Indians told him why they named the Río Verde. . . . it flows through a sierra of many veins of green, blue, and otherwise colored stones."

That is a good description of malachite copper ore. The Verde still traverses a region that has yielded great riches from Jerome's United Verde and Extension mines. The Black Hills between the river and Mingus Mountain could as well be called the Copper Hills. And yet in the same narrative, Father Sedlmayr added the contradictory statement that "one river is called Verde, owing to the verdure of the groves which adorn its banks." Those early Spaniards could not have seen the river in winter when its trees are like skeletons.

Two centuries after Kino apparently accepted the Indians' name for the river, Dr. Elliott Coues, translator and annotator of the Garcés journals, declared that some called the Verde the Río Alamos from its cottonwood borders. In *Arizona Place Names,* Will C. Barnes adds to the uncertainty by quoting Oñate, Velarde, and Coues, and then concludes that the Verde was "so-called by early Spanish explorers from the color of its water."

What should we conclude? Was it green trees, green rocks, green water, or all three that gave the Verde its name?

The laconic wording of maps and weather reports is preferable to the effusions of pretentious writers on nature. Here is how Will Barnes describes the source of the Verde: "Rises in T. 17 N., R. 2 W., west side of Chino Valley, north of Del Rio"; and then, quoting Frank Grubb of Prescott, "Breaks out of some large springs here which we consider as the actual head of the Verde River." In 1900 the city of Prescott derived its water supply from these springs.

There are various ways of coming to the grasslands of Chino Valley — north from Prescott through Granite Dells; south from Ash Fork, oriented on the east by Bill Williams Mountain, beloved of Martha Summerhayes; or my favorite way, over a rough road from Jerome, used chiefly by ranchers with pickup trucks that leave clouds of dust.

By the latter route the Verde is reached at Perkinsville. That name on the map consists of a third-generation family ranch and buildings on the river meadow, and a station on the spur line of the Santa Fe from Drake to Clarkdale, built to serve the mines at Jerome. From here the railroad enters a canyon that leads through barren hills to the Verde Valley. There is no room for a road. Walkers can use the track ties as a pathway. On its daily run, the "Verde Mix" hauls out cement from the converted smelter. The tracks are ledged along the walls above the river and blasted through tunnels. My view of it came from above when we flew down the Verde to the Salt.

The river's source lies beyond Perkinsville. From Chino Valley and Wash it flows through juniper and yucca grazing-land, eroded by arroyos bedded with blue-black rocks. By the time it reaches the ranch, the Verde has become a considerable stream. At the Perkinsville crossing, its volume in flood may be seen from the height and length of the plank-road bridge and the highwater marks on the willows. There are lush borders of watercress and animal tracks in the sand. Throughout its course the river is a vital resource for life of all kinds.

Where the stream emerges from the hills at the northwest head of the Verde Valley, Sycamore Creek contributes the first major flow of water from the Rim. Near here a sharp bend that forms a backwater gave the prehistoric pueblo of Tuzigoot its name: Apache for Crooked Water.

Built by the Sinagua along the spine of a rocky hill above the river, Tuzigoot was abandoned in the 1300s, either from drought or because of hostile tribes. Its excavation and restoration in 1934 by Edward Spicer and Louis Caywood was a benefit of the Depression. When Phelps Dodge shut down the Jerome mine and Clarkdale smelter, the labor force of several hundred Mexican nationals was shifted to the W.P.A. rolls and put to digging out Tuzigoot from the rubble of half a millennium.

As the Verde proceeds through its thirty-mile-long valley, it gathers volume from Oak, Beaver, and Clear creeks, and even more from its major tributary, the East Verde, all adding to its importance as one of Arizona's few surviving perennial rivers. The Verde is in marked contrast to the Santa Cruz and the Little Colorado, which the farther they flow the less water they have.

The presence of artesian water in a dry land is a benison unknown to dwellers in watery climes. Such is the mysterious (to non-geologists) Montezuma Well in the arid hills between the valley and the Rim. It was sketched by Maynard Dixon in 1900 on his first Arizona visit. Will Barnes described the well as follows:

Isolated limestone mesa about 100 feet above Beaver Creek. In this mesa is a huge open depression or crater about 600 feet across. Of great depth in which clear fresh water stands at all times about 75 feet above stream. There is a small opening through well at one side from which a constant stream flows into Beaver Creek. Lake always stands at same level. Water is used to irrigate adjacent fields. In the walls are a number of cliff dwellings. About 1885 a crazy man took up his abode in those caves. Was finally captured and placed in an asylum.

Cottonwoods, sycamores, alders, and willows grow where the Well's outlet emerges from a fissure in the limestone and is channeled through rock to Beaver Creek. As long ago as the twelfth century, the aboriginal peoples ditched this water to their fields. Their canal system is still to be seen. Flow of water, cress and fern, shade of trees, and the integrity of rock are the elements of a natural shrine. Here one understands why the Indians, as reported by Coronado, worshipped water.

Today the Verde Valley lacks social homogeneity. At the mouth of Oak Creek Canyon, among the dark piñons below the red-and-cream cliffs of the Rim, wealthier folk have colonized Sedona. As the land descends to the Verde, mobile homes are the favored dwelling. They are omnipresent in the working-class and retiree towns of Cottonwood and Camp Verde. While the former is comparatively law-abiding, the latter is such a rough settlement as to call for the recent transfer there, from Cottonwood, of the Yavapai County sheriff's branch office. Remnants of the interbred Yavapai Apaches live on reservations at Middle Verde and Clarkdale.

As is the case along most Arizona rivers, people have settled on the flood plains of the Verde and its creeks, learning to their sorrow what it means when the streams rise. More than once in recent years the Verde and its tributaries have washed out their beds.

The lesser streams come off the east Rim. On the west, the runoff from the Mingus range is from seasonal rains that for an hour or two turn dry arroyos into torrents. In the boom days of copper mining, the western slopes lost their forest of old junipers to the smelter furnaces. Now the dominant silva is runty mesquite. What can never be lost is the sundown light on the eastern cliffs.

Looking down from its perch on the shoulder of Mingus Mountain, the old mining town of Jerome boasts the valley's most interesting buildings, people, and social mores. Although it became a ghost town after the shutdown of the copper works, today Jerome is very much alive with productive artisans and artists. Some folks drift through while contented old-timers and passionate newcomers preserve traditions as they restore the turn-of-the-century buildings and meet in history and art groups, and give play-readings. The Cottonwood Town Library, directed by Sarah Bouquet, is one place where mountain and valley people amicably mingle in what may be seen as the cultural heart of the Verde Valley.

When viewed from the road that descends from Sedona to Cottonwood, the lights of Jerome float in space like a honeycomb island. Midway on that road, a dirt turnoff leads to a cultural oasis as vital as Jerome. This is the studio of John Waddell, Arizona's premier sculptor in figurative bronze, of which the Dance group in the Phoenix Civic Plaza is the finest example. Waddell continues an ancient respect by artists for the human form.

On the shore of Spring Creek, a perennial stream that rises in a bosque of cottonwoods and mesquites and makes a short run to Oak Creek, Waddell directs an apprentice program for young sculptors and other artists. In a great multi-story structure built by him and his helpers,

Waddell, his wife Ruth and family, and the apprentices live and work. There is also a foundry for bronze casting. It is a unique establishment, both real and ideal.

John Henry Waddell is an artist who came out of the Iowa heartland. From his birthplace in Des Moines to the Chicago Art Institute, Army service, and teaching art at Arizona State University, he went on to productive years in Mexico and Greece, and came finally to his journey's end on the edge of the Verde Valley. When his work enlisted the support of patrons and commissions for public sculptures in Phoenix, Tempe, Scottsdale, and Paradise Valley, Waddell was able to realize his vision on the Kerr ranch on Spring Creek. There he works at a magnetic center that draws creative people, famous and unknown, from far and near.

I first met Waddell at a community program in a school gymnasium. He was wearing corduroys and cape and one of the exotic hats he has acquired from around the world. Hats are his only eccentricity. He is friendly and open and free of so-called artistic temperament. He regards himself as a worker with strong and gifted hands, and blessed with the will and the imagination to free himself from academic and worldly traps. When it was recognized that the university campus in Tucson was artistically barren, Waddell was commissioned to model and cast the bronze Flute Player that now graces the approach to the University Library.

On the grass under the summer sky and to the sound of Spring Creek, friends gather with the Waddells and the apprentices to break bread at common board, while planes fly silently overhead. The Verde Valley lies beneath their skypath. Soaring at thirty thousand feet, the planes are invisible except for their vapor trails, which linger like rosy ribbons in the last light of day.

It was at the Waddells that I first met Reynold Radoccia, a young New York Italian-American architect-builder-mason, whose solar houses of stone and wood have proved me wrong when I once wrote that nothing beautiful had been built in the Verde Valley since Tuzigoot. Radoccia builds with Coconino sandstone, long a favorite of Southwest masons.

I accompanied Radoccia on a trip to buy stone at a yard in Ash Fork. Through Sedona and up Oak Creek Canyon we drove to the forested Rim south of Flagstaff. Ash Fork lies sixty miles west on Interstate 40, which I still think of as U.S. 66. It used to be on the Santa Fe mainline. To gain an easier gradient on the long haul up from the Colorado River to the Continental Divide, the line was relocated in a deep cut through the Coconino Plateau to the north.

We found the stoneyard on a spur track near the old station. There, stone from the quarries is dressed to building sizes. The flagstone comes in varying slabs of several inches thickness, which are then guillotined, sawed, and split to order. As well as for its colors, this stone is esteemed for its precise, clean fracturing. It is found in shades of red, buff, and cream, with occasional layers of lavender gray.

The foreman was at one of the quarries, a dozen miles north of town, and so we set out using our ears to guide us to him and order the several tons of stone required for Radoccia's next job. The road climbed steeply through piñons and junipers, deep with red dust made by the trucks that haul down the stone.

As we approached the mainline we heard a train, heralded by the growl of diesel engines. We waited at a crossing in the cut and it soon appeared — a long freight hauled by six smoking blue and gold Santa Fe engines, making a concerted roar that shook the earth and raised the hair on my neck. The sound of drilling led us to the quarry up a one-way road that required our four-wheel drive. Bare-chested

Right: Canyon of the Verde below Perkinsville

young Hercules were hammering, wedging, and prying out the layered flagstone, stacking and loading it on old army surplus trucks.

With samples brought from the yard, Radoccia placed his order to be trucked to Sedona, and we were on our way back through Chino Valley to Prescott and over Mingus Mountain to the valley.

Before the end of summer I saw that stone, cut into 4″ x 4″ bricks of varying lengths, rising as the walls of another solar house. Himself a working mason, Rennie was aided by two others as they cunningly varied sizes and colors of that beautiful material, all possessing the velvety sheen and fine texture of the Coconino formation.

In their way, Radoccia's houses are as beautiful as Tuzigoot, and they might stand as long. Robinson Jeffers would have liked them. I think of his words, "stones have stood for a thousand years and pained thoughts found the honey peace in old poems."

Now Rennie and his wife and daughter are planning to move to greener building fields in New Mexico. Their land and the home they have made out of stone and wood will net them a profit. The West is still to be won by those who are willing and able.

Along the Verde and its creeks there is other liquid magic than that found at Montezuma Well with its secret flow — the hydroelectric plant on the river at Childs. By water it can be reached only by raft or kayak in spring, for there is no road down-canyon. The old cavalry trail leaves Camp Verde and climbs toward the Rim. After a few miles, a road branches off to the south and threads its way thirty miles through rough country to reach a point high above the Verde at Childs. From there it switchbacks down to the river.

The water that spins the three turbines is not Verde water. Instead it is brought ten miles from the warm spring headwaters of tributary Fossil Creek, descending 1600 feet by tunnel, pipe, and flume to reach a holding lake and then fall almost vertically on the turbines below.

The Childs station was built in 1905-09 by the Arizona Power Company, primarily to supply electricity to the United Verde mine at Jerome, owned by William Andrews Clark, the Montana copper king. The other main customer was the city of Prescott, fifty miles to the west.

An illustrated article in *Electrical World* (August 11, 1910) tells the story of building this power plant in what was then a roadless wilderness. It took a labor force of 600 men — including Tonto Apaches, who also worked on Roosevelt Dam on the Salt — plus 400 mules and 150 wagons to do the job. Today this progenitor of all hydroelectric plants in Arizona is still using the original turbines to generate power for surrounding communities, including Payson, Pine, and Strawberry on the Rim, and more distant Prescott and Jerome. The plant is now part of the Arizona Public Service system.

There is a vast difference between this comparatively simple installation and the enormous ones at Boulder and Glen canyons on the Colorado. At Childs one can stand beside the turbines, deafened by the roar of water-powered machinery, then see the discharge fall into the river. At this point the Verde has gathered all but the water of the East Verde and is a wide swift stream. The canyon walls rise on either side. Up-river are the vestiges of the once popular Verde Hot Springs resort, now the burned-out haunt of transients.

The one bridge across the river between the Verde Valley and Horseshoe and Bartlett dams can be seen from kayak or airplane. It is a single-file suspension crossing over which sheep are driven between pastures in Bloody Basin and the Rim.

The Verde was first restrained in 1939 when Bartlett Dam was built by the Bureau of Reclamation as part of the Salt River Project. It was followed upstream ten years later by Horseshoe Dam. This was built by Phelps Dodge in a water tradeoff with the Salt River Project, whereby the copper company was allotted water from the upper Black River, the Salt's main fork, to supply its mine at Morenci.

These dams may be reached by roads from the west through Cave Creek and Carefree. Horseshoe is a low-level barrier with a unique self-activating spillway. Bartlett is a multiple arch dam, the largest ever built. It tilts up from the river at an angle, as beautiful in its way as a cathedral.

Ignoring a "no trespassing" sign, I drove to the top of Bartlett. From there I saw the discharge from the face of the dam, jetting far out to fall into the riverbed. Garlanded by *álamos,* the river proceeded serenely toward the Salt. Here the Verde flows with silt-free green water. It was the moment of truth when the Verde ran green — tree green, copper green, and at last water green. *¡El Río Verde verdad!*

My vision was shattered by the dusty arrival of a pickup truck. A man emerged and strode toward me. He was tall, lean, and bronzed, wearing jeans, boots, and a gun. Lassiter no less.

"What are you doing here?"

"Just looking."

"Got a permit?"

"No."

"Can't you read?"

"A bit."

"Why didn't you stop at the gatehouse?"

"Your dog didn't ask me to."

Was that a faint grin? "I work for the state," I ventured.

He laughed grimly. "The state cuts no ice up here. We're the Salt River Project."

I tried again. "I'm a fluviophile."

"A *what?*"

"A river lover."

He spat in disgust. "That's a new one."

I played my last card. "I know your project's general manager — personally, that is."

Lassiter seemed impressed. "You do?"

"I know his wife too. She's a fan of mine."

"She's a *what?*"

"She reads my articles in *Arizona Highways.*"

He seemed relieved. "So you're a writer. Why didn't you say so? I do a bit of reading myself."

From that point I got along fine with the superintendent. When I left the gatehouse, with a pat on the Doberman's head and a bundle of Project literature, he said, "Next time you see the manager, tell him he owes us a visit."

I was near the end of my Río Verde reconnaissance. Approaching union with the Río Salado, the Verde breaks out of the mountains and flows under Highway 87, the road from Mesa to Payson, Winslow, and the Hopi mesas. Through an arboreal desert, the stream goes placidly to its end, having lost all but its water and its cottonwoods.

What gave it character — its history, ancient and modern, its ruins of vanished peoples who built with stone and dug for copper, and its power to evoke the past — all this was left upstream. Here where the Verde enters the Salt above Granite Reef Dam and their blended waters are canalled away to irrigate the Valley of the Sun and serve the urban needs of the capital, the work of civilization begins.

In the mesquite jungle at the confluence, a maze of sandy trails holds no magic. Empty cans and bottles were the only artifacts. I skipped a farewell stone on the water. *Sic transit gloria fluvi.*

Río Colorado Chiquito

"ON CHARTS THEY FALL LIKE LACE," IS HOW THE GREEK ISLANDS were seen by Lawrence Durrell. On the drainage map of Arizona, the rivers that rise on Mt. Baldy appear like the spokes from the hub of a wheel. Geography as poetry.

So thick is the forest that covers the White Mountains that aerial reconnaissance fails to disclose the network of brooks and creeks that coalesce to form the rivers. The density of trees leads a newly arrived Arizonan to wonder if this can be the same state that includes the Painted Desert. Here in these mother mountains is the secret source of the life and the light that sustains Phoenix and its environs.

Water that streams off the sides of Mt. Baldy from rain and snow-melt makes three main rivers — the White, the Black, and the Little Colorado. As we have seen, the first two join to form the Salt River. Because of the lay of the land and its underlying structure, the Salt collects more water than the Little Colorado. The deep canyoned course of the Salt, as it nears the desert, made it possible to dam and store the water, thereby tempering floods, generating power, and irrigating crops, all of which create the prosperity of Phoenix and its sun-drenched valley.

This difference in volume of water and the terrain through which it flows is why the Little Colorado is one of Arizona's less important rivers. Once it has left its sources in the White Mountains, unlike the Verde and the Gila it has no major tributaries nor any significant aquifer-forming strata. Below the mountains its flow diminishes and its bed flattens out, so that without canyon walls, it has never been possible to dam the Little Colorado in the ways that those living on the Salt and the Verde made an agricultural and urban paradise.

Such development as there has been on the upper and middle reaches of the Little Colorado is due to the faith and persistence of a stubborn people, the Mormons. Beginning in the 1870s they were sent by their leaders in Utah to colonize the river from Springerville in the high country to St. Johns, Woodruff, and Holbrook.

The Mormons picked a poor river to colonize. It either ran too much or too little or not at all. Theirs is a heart-breaking history of dam after dam being swept away. Not until the state came to their aid in 1915 was it possible to build a barrier that held — Lyman Dam, between Springerville and St. Johns.

The Mormons also planted trees — the wrong trees, Lombardy poplars, a Utah favorite, which cast scant shade. Only their faith and

Left: Mountain meadow on the Little Colorado below Mt. Baldy

courage enabled them to persist and establish dynasties such as the Udalls of St. Johns.

The Little Colorado is an Arizona river *not* named by Kino. Oñate crossed it in 1604 on his way to El Mar del Sur and named it the Colorado, the Red, from its burden of colored silt. By the time Garcés came that way, nearly two centuries later, its relationship as a tributary of the big Colorado, the Grande, was known, and so he renamed it El Colorado Chiquito.

My interest in the river first came not from history or geography, but rather from reading *Vanished Arizona* and then trailing Martha Summerhayes around the state. The fording of the river at the Sunset Crossing by the Summerhayes party in April 1875 is the high point of her book.

Soon after the birth of Martha's baby at Camp Apache, Lieutenant Summerhayes was posted to Whipple Barracks at Prescott, and they set out with two ambulances of six mules each, two baggage-wagons, an escort of six cavalrymen fully armed, and a Mexican guide. When they stopped overnight at the ranch of Corydon Cooley, the celebrated scout, his having two young Apache wives shocked Martha, though not her more tolerant husband.

As they came down from the mountains through the pass that led to the river crossing, an Apache ambush was feared. Her husband provided Martha with a loaded derringer, with instructions to kill her three-months-old son and herself if the party were overwhelmed. After their safe passage, Summerhayes vented his feelings to Martha, who lay concealed with her baby on the wagon floor: "Thank God, we're out of it! Get up, Mattie! See the river yonder? We'll cross that tonight, and then we'll be out of their God d----d country!"

The river proved more perilous than the Apache threat. The ford was there at the Sunset Crossing because it was the one place in that area where a rock ledge, instead of the usual quicksand, made it possible for wagons to cross. At this time either the river was too high from the spring runoff or the guide picked the wrong place, for the party came close to disaster when the mules went under and had to be cut loose. Although the wagons were gotten across by being virtually made into boats, much of their contents was ruined, including Martha's household things and her beloved books. No longer did she have romantic associations with the poetic names of Colorado Chiquito and Sunset Crossing.

My familiarity with Little Colorado country stopped there at the Sunset Crossing and nearby Winslow. Highways and bridges have changed it all from a century ago. Vanished Arizona indeed! Twenty years ago when I trailed Martha Summerhayes around Arizona, my Chevrolet was not rugged enough for the road to the Grand Falls. From there to the Cameron bridge there was no road. The rest of the way to the confluence with the big river, the Little Colorado was inaccessibly canyoned. Years passed before the Scout enabled me to make it to the Grand Falls and the Cessna took me up to where I could see every twist and turn of the river as it snaked along to its end.

Even now with high-clearance vehicles, few people visit the Grand Falls. Because they are on the Navajo Reservation, road posting is minimal. The latest Rand McNally AAA map of Arizona does not show the falls nor the road to them, although the map issued by *Arizona Highways* does. They cannot compete with the Grand Canyon for grandeur. There is nothing there other than a desert river falling abruptly to a lower level — when it flows. That occurs only in spring or summer flood. There are no lodgings, no food, no gas, nobody.

The Grand Falls are in a desolate cinder-cone area about thirty-five miles northeast of Flagstaff. The road to them leaves the Leupp Road, which has branched off U.S. 89 — the highway to Glen Canyon Dam and Page — and then lurches a dozen miles over black sand, cinders, and lava. The river's presence in this midst of nowhere is heralded only by a low line of salt cedars.

The road suddenly ends at the brink of two wide terraces with a combined height equal to Niagara. An old lava flow created this sudden change of terrain. In spring after the runoff from melting snow in the Whites has raised the river, and in summer when flash floods from thunderstorms have made it run wild, the Little Colorado becomes a violent and dangerous stream, as the early Mormons learned all too well. When the wind is strong, the chocolate-colored silt is blown off the falls as dust. The sight and sound are awesome, the more so in that almost no one is ever there. The Navajos have furnished a few ramadas, picnic tables, and privies, although there is no drinking water nor a railing along the edge. Not long ago an unsuspecting viewer was blown to his death. It is no place for children unless they are part mountain goat.

From Cameron, if the river were *not* running (which is most of the year), one could walk down the dry bed to the confluence, although sudden floods and quicksand would quickly make it dangerous. The gorge is deep and narrow with vertical walls. Along the road are points where one can park and walk to the brink. It was frustrating not to be able to gain the riverbed. And so I took to the air again.

My rendezvous with Chris and Michael was in the Verde Valley at the Cottonwood airport on a morning in July. It is not a major airport. According to the *Verde Independent,* the weekly Voice of the Valley, the town manager recently requested the sheriff of Yavapai County to keep stray cattle off the runway. It would help, that officer's minion replied, if the town would keep the approaches mowed.

I was early and kept scanning the sky for the plane, due on the brief flight from Flagstaff. The tips of the Peaks showed above the Rim. As it came in for landing, the Cessna looked like one of the eagles that patrol the valley.

The plane taxied up and my two young colleagues climbed down and strode toward me. Again I thought how much they resemble mountain men of the time of Old Bill Williams. Whereas Michael is tall and lanky and long-haired, Chris is short and black-bearded, with a wrestler's chest and shoulders — and an old gray hat that he swears to wear until he completes his doctorate at the University of New Mexico.

"Howdy, old-timer," Michael said. "How's your stomach today? Pretty rough up there."

"No problem," was my cool answer. "I left it home."

They grinned. "Tell us some more of your adventures in small planes," Chris teased.

And so, over a Pepsi, I recounted how once on a charter flight from Bozeman to Butte, the pilot landed his Piper in the Lewis and Clark Cave and took off again. And to impress them even more, how as a boy I had built a six-foot kite that once took me on a short flight.

Up and in I climbed and to Michael's "Prop clear!" Chris turned it over and we were swiftly airborne, I on the bench seat with the usual load of maps, tools, thermos, and fig bars. Beneath us passed the green banks of the Verde, Tuzigoot, the white-roofed Waddell studio, rock-red Sedona, Oak Creek, the forest, and Flagstaff, all in a matter of minutes.

Mountains were on either side as we flew northwest to the Colorado — Bill Williams on the left, the Peaks on the right. Ponderosas thinned to juniper and piñon. The road was beetled with cars on their way to the Grand Canyon.

We reached the canyon at El Tovar and I saw the brown wooden hotel whose halls and stairs I had roamed as a child. "The wise writer

Following pages: Gorge of the Little Colorado below Cameron;
Confluence of the two Colorados

47